Electron diffraction and optical diffraction techniques

Practical Methods in
ELECTRON MICROSCOPY

Edited by

AUDREY M. GLAUERT

Strangeways Research Laboratory

Cambridge

1973

NORTH-HOLLAND PUBLISHING COMPANY – AMSTERDAM · LONDON

AMERICAN ELSEVIER PUBLISHING CO., INC. – NEW YORK

ELECTRON DIFFRACTION AND OPTICAL DIFFRACTION TECHNIQUES

B. E. P. BEESTON, B.A., M.S., Ph.D.

AEI Scientific Apparatus Ltd.,
Harlow, Essex

ROBERT W. HORNE and ROY MARKHAM

John Innes Institute,
Colney Lane, Norwich

1973
NORTH-HOLLAND PUBLISHING COMPANY – AMSTERDAM · LONDON
AMERICAN ELSEVIER PUBLISHING CO., INC. – NEW YORK

Library of Congress Catalog Card Number 72–88033
North-Holland ISBN for the series 0 7204 4250 8
North-Holland ISBN for this book 0 7204 4253 2
American Elsevier ISBN 0 444 10411 9

Publishers:

NORTH-HOLLAND PUBLISHING COMPANY – AMSTERDAM
NORTH-HOLLAND PUBLISHING COMPANY, LTD. – LONDON

Sole Distributors for the U.S.A. and Canada:
AMERICAN ELSEVIER PUBLISHING COMPANY, INC.
52 VANDERBILT AVENUE
NEW YORK, N.Y. 10017

This book is the laboratory-edition of Volume 1, Part II, of the series 'Practical Methods in Electron Microscopy'.
Volume 1 of the series contains the following parts:
Part I Specimen preparation in materials science
by P. J. Goodhew
Part II Electron diffraction and optical diffraction techniques
by B. E. P. Beeston,
Robert W. Horne and Roy Markham

Printed in The Netherlands

Contents

ERRATA

PRACTICAL METHODS IN ELECTRON MICROSCOPY
Volume 1

Edited by:
AUDREY M. GLAUERT.

1) On page 236, line 16 should read 'for the f.c.c. lattice h, k and l are ALL EVEN or ALL ODD.'

2) On page 248, line 9 should read '(i) the f.c.c. structure h, k and l must be all even or all odd.'

3) On page 249, line 10 should read '(411) or (330) 18'

2 APPLICATION OF OPTICAL DIFFRACTION AND IMAGE RECONSTRUCTION TECHNIQUES TO ELECTRON MICROGRAPHS

BY ROBERT W. HORNE AND ROY MARKHAM

1 AN INTRODUCTION TO ELECTRON DIFFRACTION

B. E. P. BEESTON, B.A., M.S., Ph.D.
AEI Scientific Apparatus Ltd.,
Harlow, Essex

Acknowledgements

I would like to thank the many friends and colleagues who have assisted in the preparation of this book, both with invaluable discussions and encouragement in moments of difficulty.

I am indebted to Mr. J. F. Hamilton, Research Laboratories, Eastman Kodak Co., Rochester, N.Y. for supplying the original print from which Fig. 3.7 was made, and also to Mr. T. Gridley, 83 Chalkdell Fields, St. Albans, Herts, for producing all the line diagrams from my rough sketches.

Finally I wish to thank AEI Scientific Apparatus Ltd., for the provision of laboratory facilities and for encouraging the publication of this text.

June 1972 B. E. P. BEESTON

Chapter 1

What is an electron diffraction pattern?

Whenever an image is formed on the viewing screen or photographic plate in the electron microscope, an electron diffraction pattern is present in the back focal plane of the objective lens of the instrument. This is the case regardless of whether the specimen is amorphous or crystalline, tissue or metal, synthetic or natural. If the specimen is amorphous, that is the constituent atoms or molecules are arranged entirely randomly and without any distinct repeating structure, then the diffraction pattern will contain no distinct maxima and will consist only of diffuse scattering around the bright central spot. Such a diffraction pattern is shown in Fig. 1.1, taken from a thin carbon film. Other specimens which are amorphous and therefore give only diffuse electron diffraction patterns are, (a) much biological tissue, (b) glass-like materials and (c) some evaporated thin films of metals and other materials.

The electron diffraction pattern is essentially a record of the structure of the specimen under examination, and since an amorphous material contains no periodic structure, it follows that its diffraction pattern will contain very little information, except of course, that the specimen is in fact amorphous, and not crystalline. However many specimens are crystalline, that is their constituent atoms, molecules or clusters of molecules are arranged in a distinct repeating array, in two or three dimensions. For purposes of simplicity reference will only be made to arrays made up of atoms, but it must be remembered that structures also exist that are made up of repeating arrays of molecules or clusters of molecules.

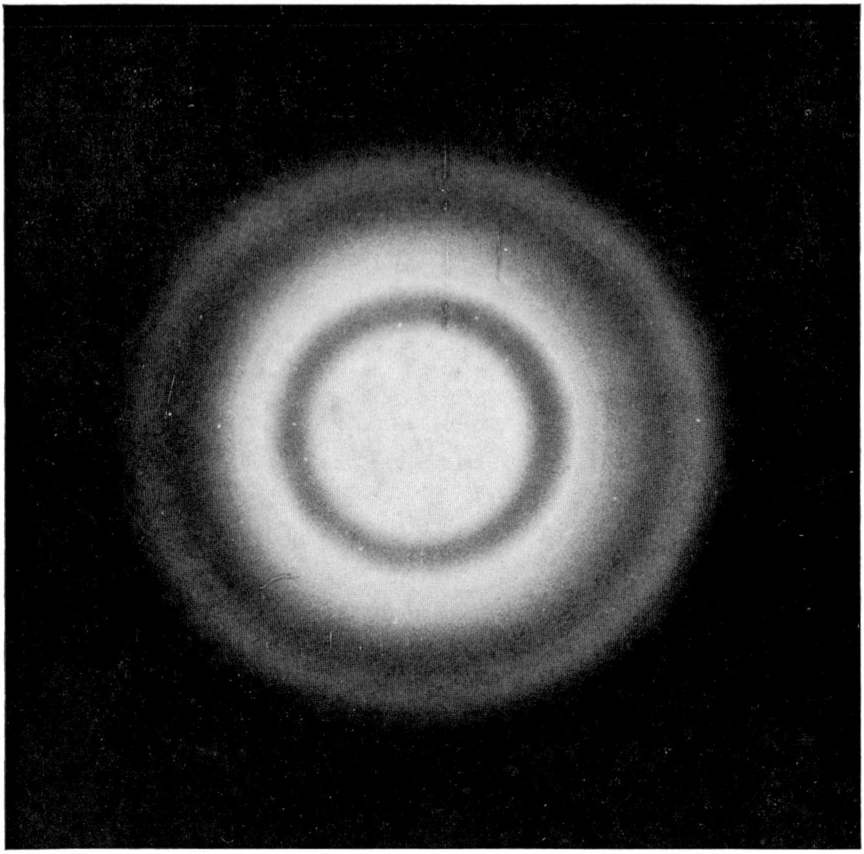

Fig. 1.1 Electron diffraction pattern from a thin carbon film, showing the diffuse scattering around the bright central spot and no distinct diffraction maxima.

1.1 Single crystal diffraction pattern

The diffraction pattern from a crystalline specimen is a record of the periodic structure or repeating array of that specimen. If the specimen is a *single crystal*, that is a single repeating array of atoms, such as in Fig. 1.2, extending in the third dimension into the paper, then the diffraction pattern will consist of a simple array of bright spots, as in Fig. 1.3. This is known as a single crystal diffraction pattern. The arrangement of the spots in the single crystal diffraction pattern depends upon the orientation of the atomic array in the electron microscope. If, for instance, four arrays of atoms, identical to Fig. 1.2, are stacked one on top of each other, making a cube with four atoms

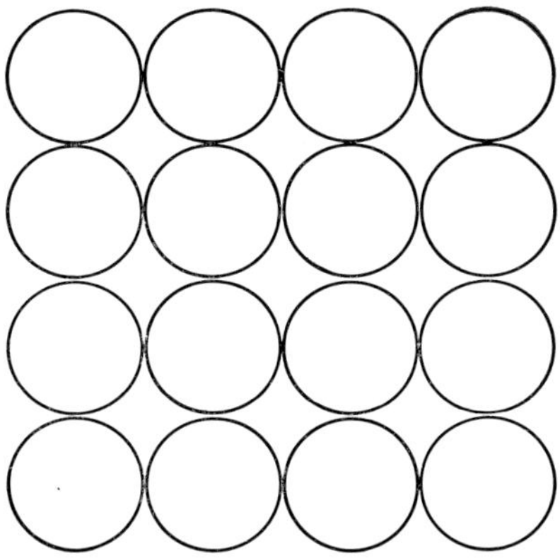

Fig. 1.2. Single crystal square array of atoms or molecules.

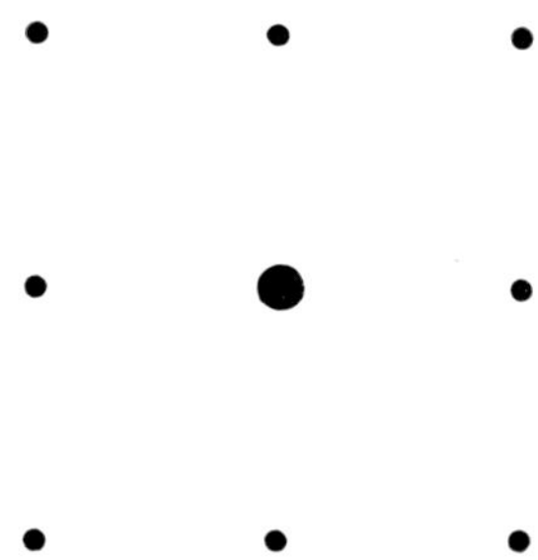

Fig. 1.3. Single crystal electron diffraction pattern of a square array of atoms or molecules, such as in Fig. 1.2 (schematic).

Fig. 1.4. Hexagonal appearance of a cubic array of atoms when viewed directly on to one of the corners of the cube.

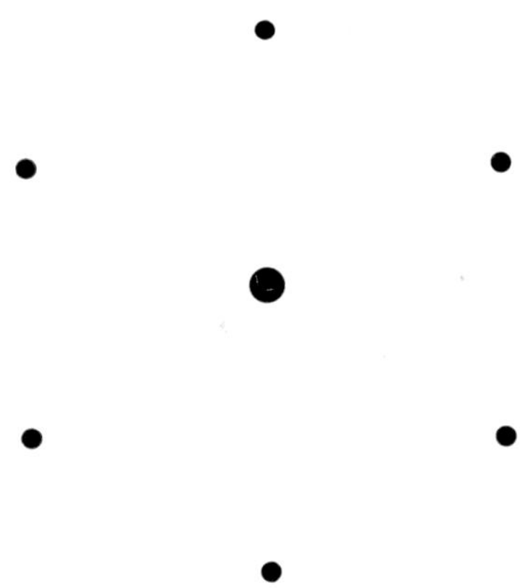

Fig. 1.5. Electron diffraction pattern from a cubic array of atoms orientated in the microscope as in fig. 1.4 (schematic).

along each edge, and if this cube is viewed directly on to one of its corners, then the atomic array will appear as in Fig. 1.4, and the corresponding electron diffraction pattern will have an hexagonal array of discrete diffraction spots, (Fig. 1.5). Thus the diffraction pattern of an array of atoms may be considerably altered by viewing that array at different orientations.

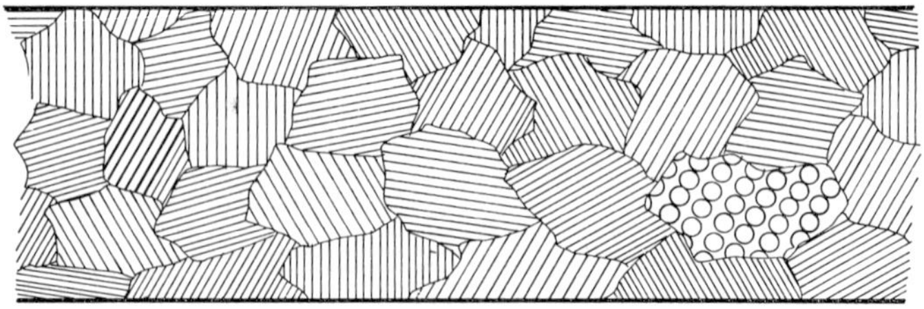

Fig. 1.6. Schematic polycrystalline specimen containing many separate grains with identical atomic arrays at different orientations.

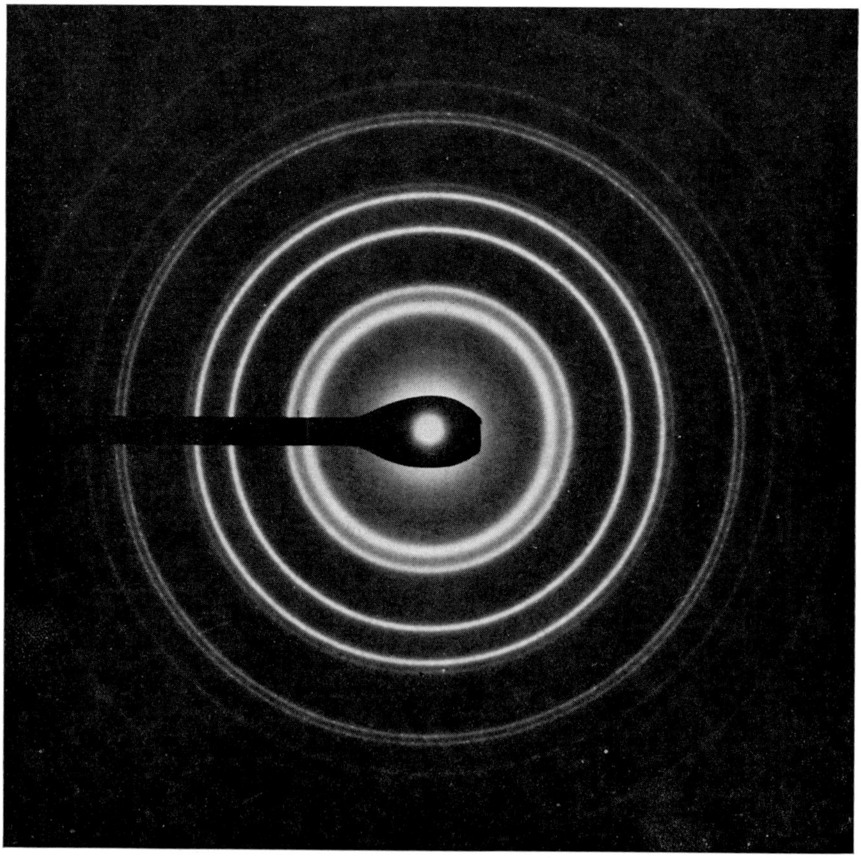

Fig. 1.7. Polycrystalline diffraction pattern from a specimen such as Fig. 1.6.

1.2 Polycrystalline diffraction pattern

If the specimen consists of a large number of small discrete areas, each with exactly the same atomic array, but at different orientations to each other (Fig. 1.6), the specimen is termed *polycrystalline*, and its diffraction pattern will then consist of a series of concentric rings (Fig. 1.7). It can easily be seen that this pattern is really a large number of single crystal diffraction patterns, each rotated by a small amount with respect to each other; the same pattern might be formed by taking a long time exposure, while a single crystal was continuously rotated. Additional rings appear for all the other diffraction patterns which occur by viewing the atomic array in different orientations. Thus a single crystal diffraction pattern is a special case of the general polycrystalline pattern, and contains only part of the total structural information which may be obtained from the crystal by electron diffraction analysis.

There are also numerous intermediate diffraction patterns between these two extremes. The specimen may consist of only two crystals, perhaps in a specific orientation with respect to each other. This is the case for twins in certain metals and minerals, when one crystal is the mirror image of the other, and the diffraction pattern contains two superimposed patterns, rotated with respect to each other, as discussed in § 7.5.

1.3 Preferred orientation diffraction pattern

A specimen may contain a large number of discrete areas, generally known as *grains*, each with the same repeating or periodic array, which are only partially randomly oriented, in that they prefer where possible to lie in, or close to, a particular orientation. Such a specimen is said to contain *preferred orientation*, and may occur, for instance, when a metal rod, in which the grains are initially randomly oriented, is drawn through a die to form a thin wire, when many of the grains prefer to line up close to the direction of drawing. In this case, the diffraction pattern will consist of concentric rings, but with certain parts of the rings very much brighter than others, representing the preponderance of grains lying in certain orientations (Fig. 1.8).

Thus it is clear that there is a close qualitative link between the arrangement of the diffraction pattern and the structure of the specimen under observation. In addition, there is a close quantitative link between the two, which is the basis of electron diffraction analysis, and is described in detail in the following chapters.

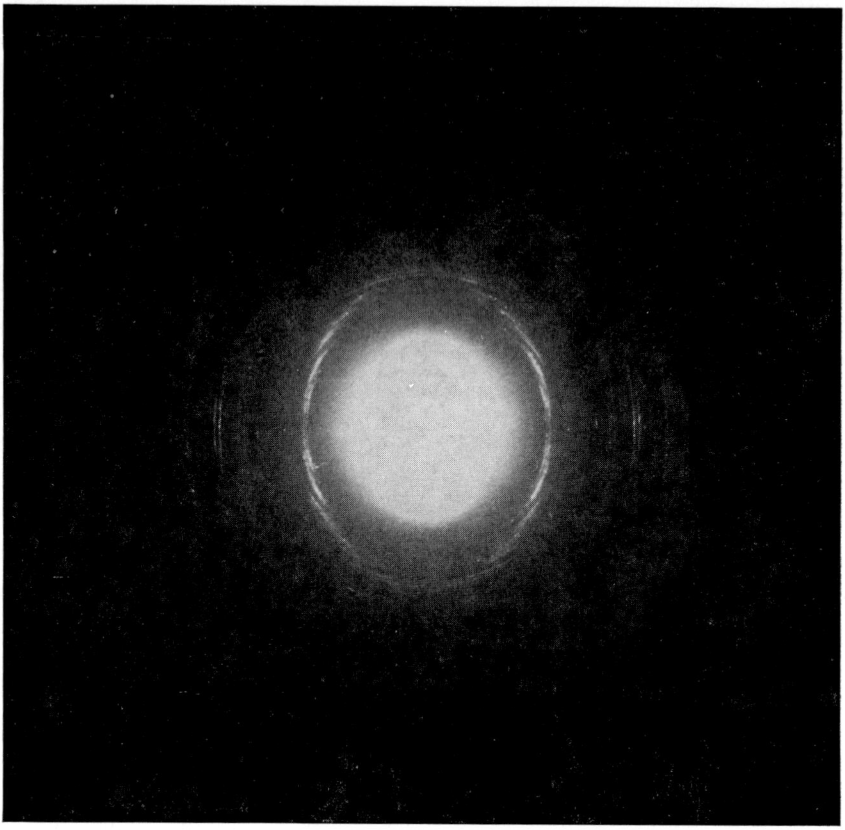

Fig. 1.8. Preferred orientation electron diffraction pattern.

Chapter 2

Information contained in the electron diffraction pattern

In Chapter 1 it was shown that the electron diffraction pattern contains information about the structure of the specimen under examination. In order to determine this structural information quantitatively, it is necessary to know how the structure of the specimen is defined. This involves some knowledge of crystallography, and a simple treatment is given here, with further data and diagrams in Appendices 1 and 2.

2.1 Crystallography in two dimensions

Initially consider only a two dimensional array of atoms. The simplest would be a square array, as in Fig. 1.2. This diagram assumes that the distances between the centres of adjacent atoms in the major directions in the array are equal, but this is often not the case. To overcome the difficulty of representing this diagramatically, the atoms are usually represented by their centre points only joined by lines representing their distances apart. Fig. 2.1a illustrates a square array in which the atoms are equidistant along the major axes, x and y, while Fig. 2.1b illustrates a rectangular array with different *spacings* between the atoms along the two major axes.

2.2 Crystallography in three dimensions

Crystalline specimens rarely have a regular array in only two dimensions. Some specimens, in particular protein molecules, such as ferritin, and some virus particles, deposited on a carbon support film, do arrange themselves

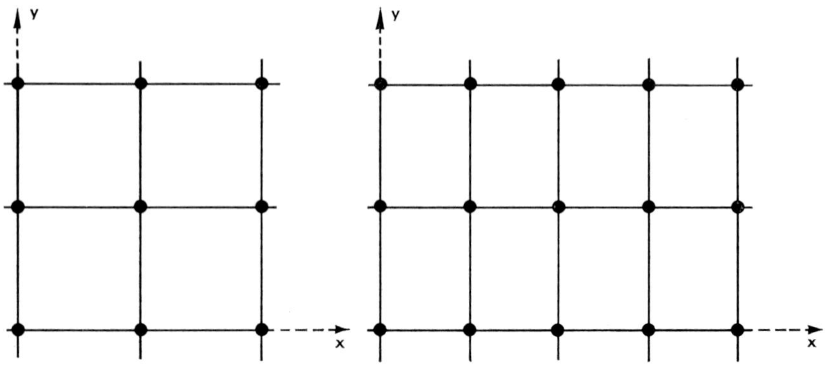

Fig. 2.1a   Fig. 2.1b

Square and rectangular arrays of atom sites, in which the atoms are represented by points and the interatomic spacings by bars.

in a two-dimensional array, but in general the model must be extended into three dimensions.

2.2.1 CRYSTAL SYSTEMS

Consider first a *simple cubic* array, or *lattice* as it is generally known, in which the atomic spacings are the same along the three major axes, the x, y and z axes. This lattice is shown in Fig. 2.2, in which the solid dots at the corners

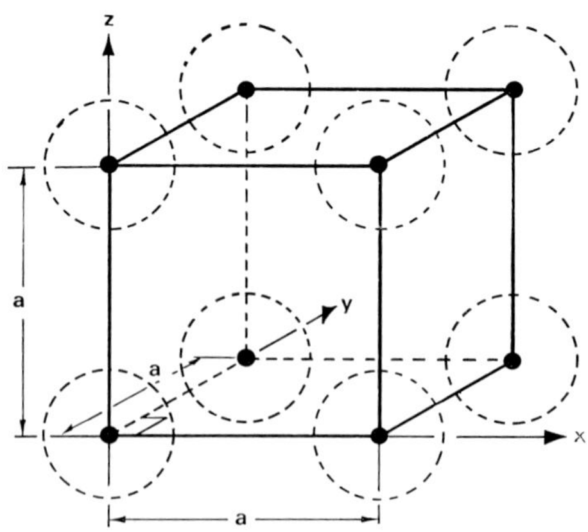

Fig. 2.2. Cubic array of atom sites, with points representing the individual atoms at the corners of the cubic unit cell. The edges of the unit cell along the three major axes are all equal and termed a, the lattice parameter.

of the cube represent the individual atoms making up the atomic array. The lattice spacings, being identical along the x, y and z axes, are all termed a, the *lattice parameter*. Each crystal lattice is represented by its *unit cell*, that is, the smallest 'building block' of the lattice which may be drawn, while still retaining the structural arrangement of the constituent atoms.

Now if the spacings along one of the axes, say the z axis, is different from that along the other two, the angles between the axes remaining 90°, the structure is termed *tetragonal*, and the lattice spacings along the x, y and z axes are termed a, a and c respectively, as in Fig. 2.3. Further, if the spacings along all the three axes are different, while the angles between the axes remain 90°, the structure is termed *orthorhombic*, and the lattice spacings along the x, y and z axes are termed a, b and c, respectively, as in Fig. 2.4.

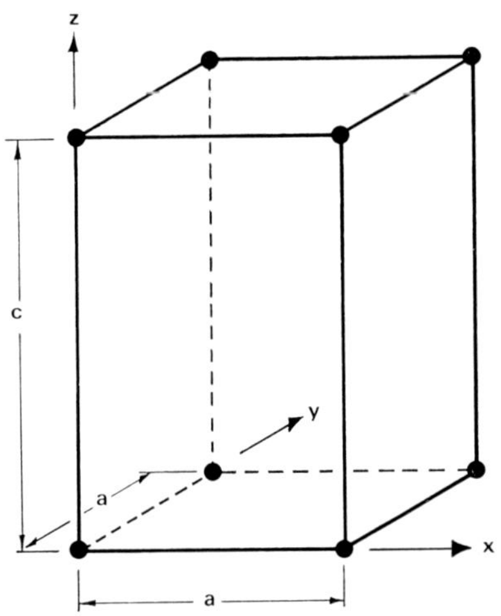

Fig. 2.3. Tetragonal array of atom sites, in which the unit cell lengths along the x and y axes are the same, and termed a, and different from that along the z axis, termed c.

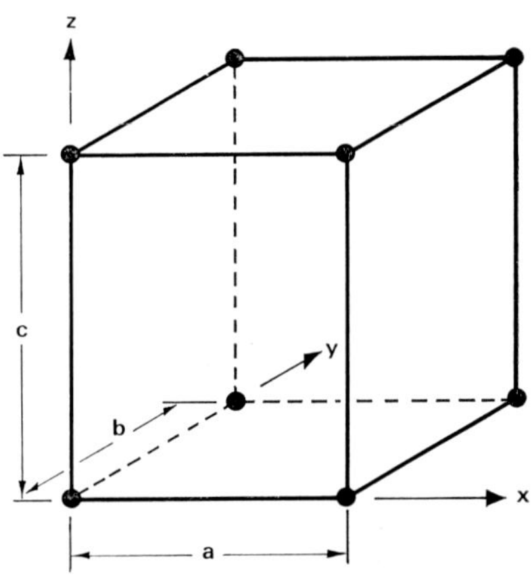

Fig. 2.4. Orthorhombic array of atom sites, in which the unit cell lengths along the three major axes are all different and termed a, b and c.

There are other simple lattices formed when the angles between the major axes, defined as

α between the y and z axes
β between the x and z axes
γ between the x and y axes

as in Fig. 2.5, are no longer restricted to 90°. Seven major lattices, or *crystal systems* as they are called, can be formed by various combinations of the

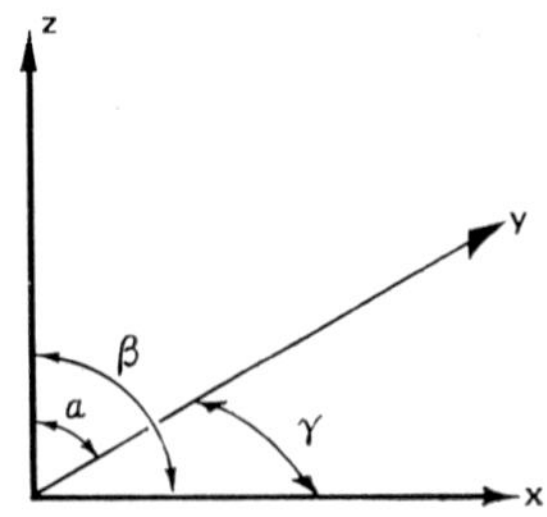

Fig. 2.5. Angles α, β and γ between the three major axes of the unit cell, x, y and z.

TABLE 2.1

The seven crystal systems

Crystal system	*Lattice spacings*	*Angles between axes*
Cubic	$a = b = c$	$\alpha = \beta = \gamma = 90°$
Tetragonal	$a = b \neq c$	$\alpha = \beta = \gamma = 90°$
Orthorhombic	$a \neq b \neq c$	$\alpha = \beta = \gamma = 90°$
Rhombohedral (trigonal)	$a = b = c$	$\alpha = \beta = \gamma < 120° \neq 90°$
Hexagonal	$a = b \neq c$	$\alpha = \beta = 90°; \gamma = 120°$
Monclinic	$a \neq b \neq c$	$\alpha = \gamma = 90° \neq \beta$
Triclinic	$a \neq b \neq c$	$\alpha \neq \beta \neq \gamma \neq 90°$

parameters a, b and c and α, β and γ, as shown in Table 2.1. These are the seven crystal systems, into one of which any crystalline structure must fall.

2.2.2 CRYSTAL LATTICES

There are a number of standard variations of these crystal systems, as for instance when there is an atom at the centre of the unit cell in the cubic system (Fig. 2.6a). Also, there may be atoms at the centres of all the faces in the unit cell of the cubic system (Fig. 2.6b). These variations, or *crystal*

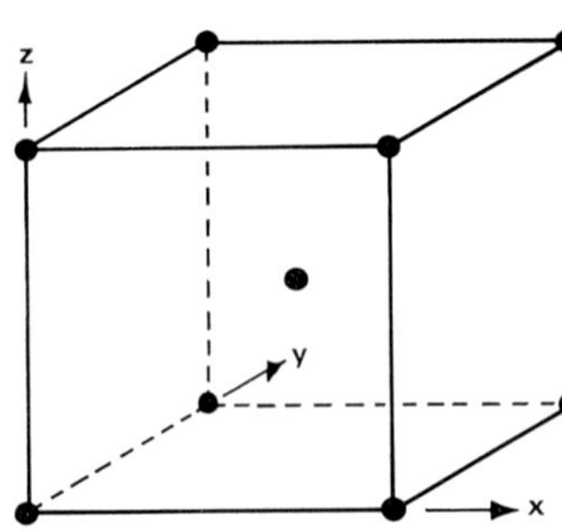

Fig. 2.6a. Body-centred cubic unit cell.

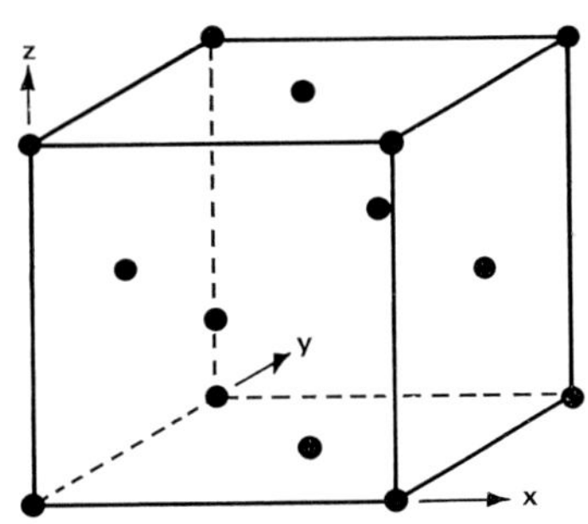

Fig. 2.6b. Face-centred cubic unit cell.

lattices as they are termed, are known as *body-centred cubic* (b.c.c.), and *face-centred cubic* (f.c.c.), while the simple unit cell is termed *primitive* or *simple cubic*. Similar crystal lattices exist in some of the other crystal systems, although fewer in those of lower *symmetry*, that is with fewer mirror planes, so that the triclinic system exists only as the primitive lattice. There are fourteen crystal lattices in all, as shown in Table 2.2,

TABLE 2.2

The fourteen crystal lattices

Crystal system	*Crystal lattices*	
Cubic	p., b.c., f.c.	P, I, F
Tetragonal	p., b.c.	P, I
Orthorhombic	p., b.c., f.c., o.f.c.	P, I, F, C
Rhombohedral (trigonal)	p.	P, (R)
Hexagonal	o.f.c.	C
Monoclinic	p., o.f.c.	P, C
Triclinic	p.	P

where p. or P represents primitive
b.c. or I represents body-centred
f.c. or F represents face-centred
and o.f.c. or C represents one face-centred.

All these crystal lattices are illustrated in Appendix 1, but an intimate knowledge of them is not necessary at this point; it suffices to appreciate that they exist and that all crystal structures may be represented by one of these crystal lattices.

2.2.3 LATTICE PLANES

Now returning to the crystal lattice itself, remember that it is defined by the lattice spacings along the three major axes, *a*, *b* and *c*, or *lattice parameters* as they are known, and the angles between the axes, α, β and γ. The crystal lattice has so far been considered as made up of a number of points, which are at the centres of the atoms, but for many purposes it is more useful to think of the planes that can be drawn joining the lattice points rather than the points themselves. Thus, in Fig. 2.7, it is preferable to think of planes

ABCO drawn through points A, B, C and O
ABHG drawn through points A, B, H and G
and CBHE drawn through points C, B, H and E

rather than the lattice points themselves.

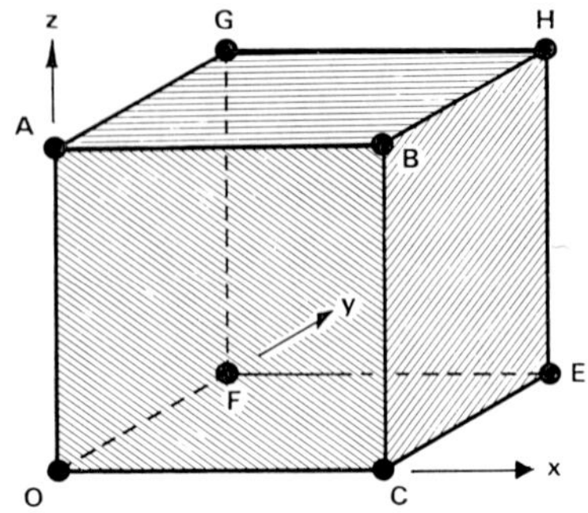

Fig. 2.7. Simple cubic unit cell showing the cube faces as lattice planes joining the individual atom sites.

2.2.4 IDENTIFYING THE LATTICE PLANES – MILLER INDICES

The system of identifying lattice planes used in § 2.2.3 is far from satisfactory in practice, particularly as the positioning of the origin, O, of the lattice at the front left hand corner of the cube is entirely arbitrary, and plane ABCO is identical to GHEF, as is CBHE to OAGF, and ABHG to OCEF. In order to overcome this difficulty a system of indices, known as Miller indices, has been universally adopted. The indices (h, k and l) are defined as the reciprocals of the intercepts that any particular plane makes along the three major axes, in terms of the lattice parameters a, b and c. Thus for a plane AFC, in Fig. 2.8, that intercepts the x, y and z axes at one lattice parameter distance along each axis from the origin O, the intercepts are

$a/1$ along the x axis
$a/1$ along the y axis
$a/1$ along the z axis

so that the Miller indices for plane AFC are (111).

A simple plane, such as CBHE, which cuts the x axis at one lattice parameter distance, yet lies parallel to the y and z axes, has intercepts

$a/1$, $a/0$ and $a/0$, along the x, y and z axes,

and so has Miller indices (100). Likewise,

plane FGHE is denoted (010)
and plane ABHG is denoted (001).

Now any plane that is exactly parallel to an indexed plane and has the same interplanar spacing, is denoted by the same indices. Thus plane OAGF is also

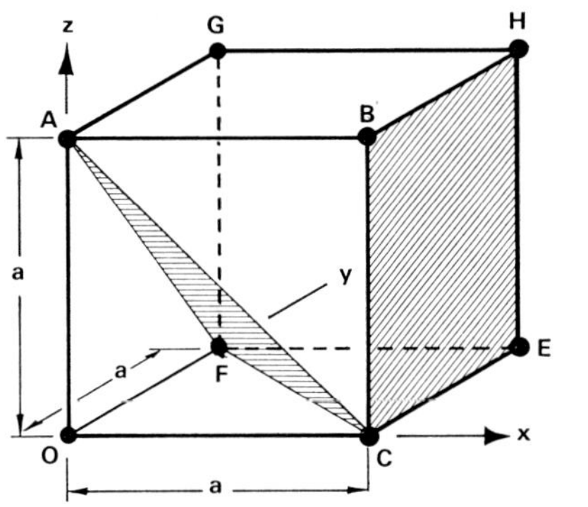

Fig. 2.8. Simple cubic unit cell showing the (111) and (100) lattice planes.

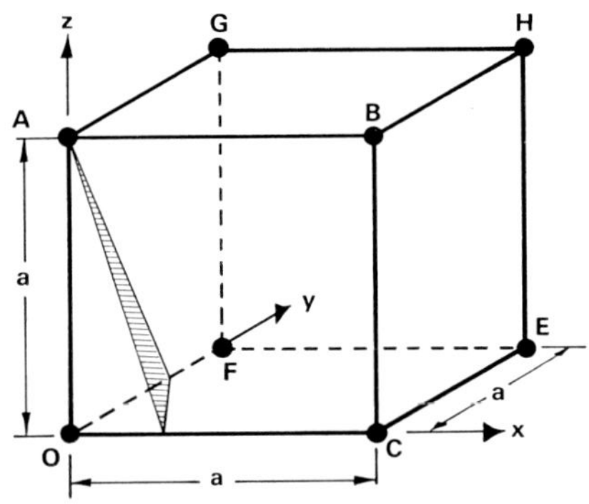

Fig. 2.9. Simple cubic unit cell showing the (321) lattice plane.

indexed (100), where ( ) are used to denote a particular set of planes. Curly brackets { } are used to indicate a group of planes of the same type for instance (100), (010) and (001) are all of the type {100}. A complex plane which cuts the major axes, for instance, at 1/3, 1/2 and the full distance along the three major axes of the unit cell, as in Fig. 2.9, will be indexed (321).

2.2.5 SPACINGS BETWEEN THE LATTICE PLANES – *d*-SPACINGS

Now, of course, these planes are not simply abstract planes joining arbitrary points along the major axes of the unit cell, but these are the planes which

Fig. 2.10. Parallel sets of lattice lines drawn in a two-dimensional array of atoms.

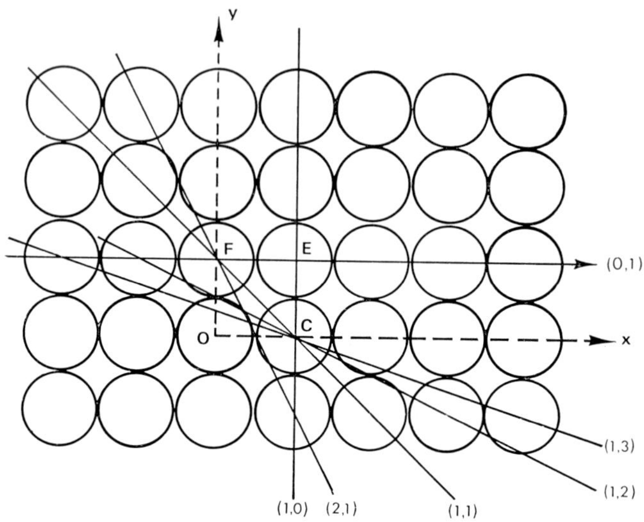

Fig. 2.11. Lattice lines drawn in a two-dimensional array with only one line representing each set of parallel lines. The lines are indexed in terms of the reciprocals of the intercepts along the major axes.

actually contain the atoms which make up the crystalline lattice. To clarify this, consider for a moment the original two-dimensional array (Fig. 1.2). From Fig. 2.10, it is clear that a large number of lines may be drawn which contain a number of atoms. If for simplicity only one line is drawn to represent each of the sets of parallel lines, as in Fig. 2.11, it is clear that there are many possible lines which can be drawn joining a greater or fewer number of atoms. By inspection it is possible to index each line, in terms of the reciprocal of its intercept along the two major axes, x and y, of the unit array.

The third dimension is easily added to the array, making a three-dimensional lattice, by placing a number of arrays identical to that in Fig. 2.11, one on top of the other extending out of the page. The array OCEF now becomes the basal plane of the unit cell of Fig. 2.2. A number of planes can be drawn joining large numbers of atoms in this lattice, each plane being indexed with Miller indices as previously indicated. These planes are often drawn without indicating the atoms they contain, as in Fig. 2.12a, or even as in Fig. 2.12b, where it is understood that the third dimension is perpendicular to the plane of the page.

It is the spacings between the individual planes, in these sets of parallel planes, known as *d-spacings*, and the orientation relationships between the various sets of parallel planes, that can be determined by electron diffraction

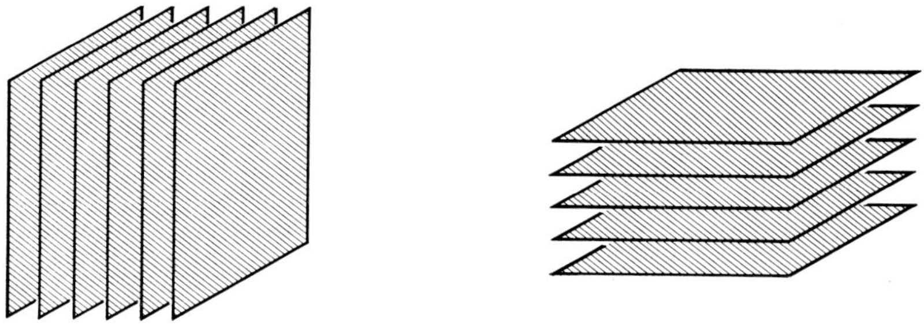

Fig. 2.12a. Sets of parallel lattice planes drawn without indicating the individual atoms they contain.

Fig. 2.12b. Sets of parallel lattice planes drawn as lines where it is understood that the planes extend perpendicular to the plane of the page.

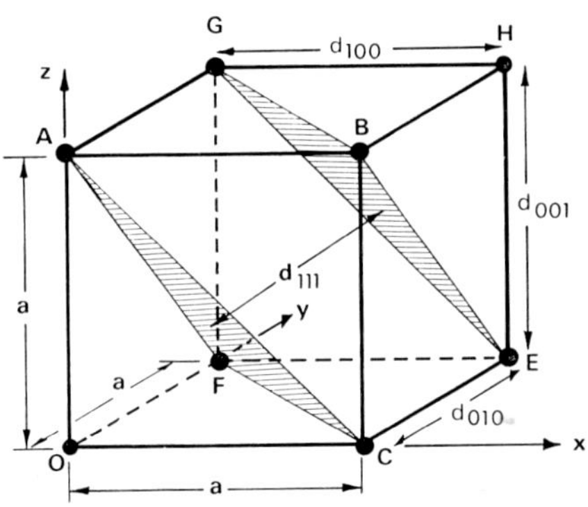

Fig. 2.13. Simple cubic unit cell showing the d_{100}, d_{010}, d_{001} and d_{111} lattice plane spacings.

analysis. To indicate the set of planes it is for which the *d*-spacing is being measured, the symbol *d*, is generally given the suffix *hkl*, such that

d_{100} indicates the spacing between (100) planes, e.g. OAGF and CBHE, in Fig. 2.8,

d_{010} indicates the spacing between (010) planes, e.g. OABC and FGHE, and

d_{001} indicates the spacing between (001) planes, e.g. OCEF and ABHG.

For the cubic lattice in Fig. 2.13, $d_{100} = d_{010} = d_{001} = a$, the lattice parameter. Other sets of planes will have smaller spacings, for example the spacing d_{111}, between (111) planes, eg. AFC and GBE, is found by geometry to be, $d_{111} = a/\sqrt{3}$. There is a general formula for the *d*-spacings, in terms of the lattice parameters, *a*, *b* and *c*, the *h*, *k* and *l* values and the α, β and γ values, for each of the seven crystal systems.

For the cubic system, in which $a = b = c$ and $\alpha = \beta = \gamma = 90°$,

$$d_{hkl} = \frac{a}{\sqrt{(h^2 + k^2 + l^2)}}$$

so that

$$d_{111} = \frac{a}{\sqrt{(1 + 1 + 1)}} = a/\sqrt{3}$$

The *d*-spacing formulae for the other crystal systems are given in Appendix 2a. It is these d_{hkl} spacings that are the basic information that can be obtained from the electron diffraction pattern.

2.3 Information obtainable from the electron diffraction pattern

Once the *d*-spacings have been determined for the specimen under examination, there are a number of other pieces of information that can be obtained, some requiring other corroborative information, some a prior knowledge of the structure or elemental analysis of the specimen, and some a certain amount of intuition. For a completely unknown material, it should be possible to determine

(a) the crystal lattice to which the specimen belongs, and

(b) the lattice parameters,

and if some prior knowledge of the elemental analysis of the material is available, or if from a knowledge of the specimen history there are only a certain limited number of materials which may constitute the particular specimen, it may be possible to determine

(c) the exact composition of the unknown material.

Such an analysis is considerably aided by the comprehensive tabulations of the *d*-spacings of very many elements, alloys and mineral compounds, determined by previous X-ray investigations (A.S.T.M. Index). Since the *d*-spacings are almost invariably unique to particular materials, and are, of course, independent of the technique (X-ray or electron diffraction), by which they are measured, a search through the tabulated list of materials for a set of *d*-spacings to match those of the unknown specimen may lead to a unique analysis. Of course the searching time is much reduced if some prior indication of the possible elemental constituents is available.

Measurements from the diffraction pattern may not be very accurate, and this may allow a number of materials to fit the measured *d*-spacings within the limits of the possible errors. Again, if by prior X-ray microanalysis a knowledge of the elements present is available, or if there are only a few possible alternatives that the unknown material can be, it may be possible to eliminate all but one of the possibilities.

Frequently the composition of the material under examination is well known, and then the electron diffraction pattern can be used to determine

(d) the orientation in which the specimen is lying with respect to the incident electron beam, if a single crystal is being examined, or

(e) the orientation relationship between two or possibly a larger number of single crystals, if it is a polycrystalline specimen, or

(f) the orientation relationship between precipitates, inclusions or twins and the surrounding matrix, or

(g) the crystallographic orientation of defects of all kinds, dislocations, stacking faults, voids, boundaries ... in the matrix, by correlating the image and the diffraction pattern of the same area of the specimen.

Clearly there is much information which can be extracted from the specimen, if it is possible to determine its *d*-spacings from the available diffraction patterns. This is obviously the main objective, but first it is necessary to describe how the electron diffraction pattern is formed in the electron microscope and how it is observed and recorded.

FURTHER READING

Barrett, C. S. (1952), Structure of metals (McGraw Hill Comp. Inc., New York).
Cohen, J. B. (1966), Diffraction methods in materials science (The Macmillan Comp., New York).
Cullity, B. D. (1962), Elements of X-ray diffraction (Addison-Wesley Publ. Comp. Inc., Reading, Mass.).
Guinier, A. (1952), X-ray crystallographic technology (Hilger and Watts, Ltd., London).
Kittel, C. (1968), Introduction to solid state physics (J. Wiley and Sons, Inc., New York).
Quinn, T. F. J. (1970), X-rays, electrons and crystalline materials (Butterworth and Co. Ltd., London).
Sands, D. E. (1969), Introduction to crystallography (W. A. Benjamin, Inc., New York).

Chapter 3

Formation of the electron diffraction pattern

3.1 Scattering of the electrons by the specimen

As the beam of electrons passes through the specimen in the microscope, some of the electrons are deflected or scattered from the main beam, in various directions and at various angles. The denser the material and the thicker the specimen, the greater the proportion of the incident electrons that will be scattered, and the larger the angles at which these electrons will be scattered. This is shown schematically in Fig. 3.1, for an amorphous specimen of varying density and thickness, together with the corresponding intensity distributions, and the diffraction patterns that will be observed on the final viewing screen, Thus even for amorphous materials the diffraction pattern is related to the character of the specimen although the diffraction patterns are very diffuse and contain little information that can be readily analysed quantitatively.

However, if the specimen is crystalline, peaks of intensity occur in the diffraction pattern, in the form of concentric rings if the specimen is polycrystalline (Fig. 1.7), or as regularly spaced diffraction spots if the specimen is a single crystal (Fig. 1.3), or contains only a few small crystals. These diffraction maxima appear because, in addition to some diffuse scattering around the central spot, as the electrons pass through a crystalline specimen only those electron 'waves' which are scattered from successive parallel lattice planes lying at particular orientations, through the depth of the crystal, are 'in phase' and combine constructively. Electron 'waves' scattered from lattice planes lying at all other orientations will be 'out of phase' and

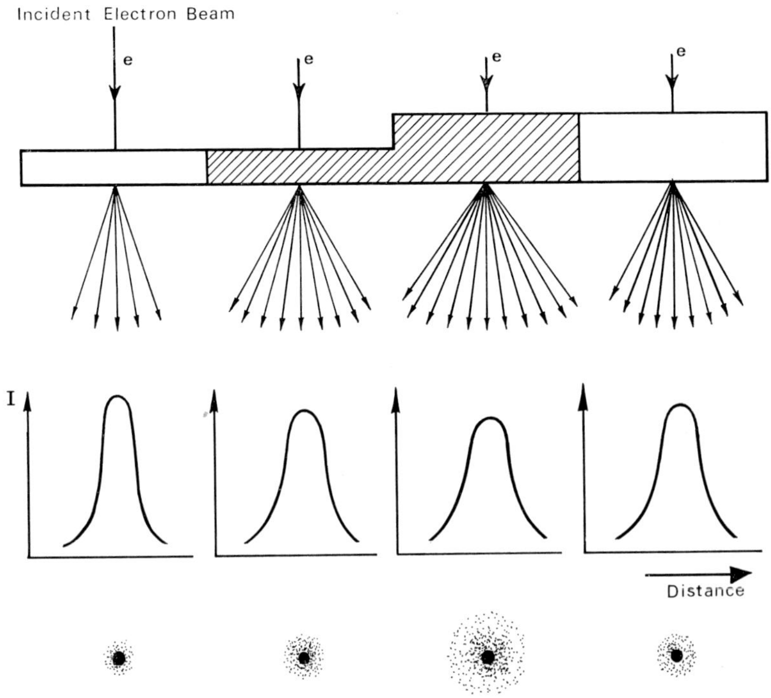

Fig. 3.1. Schematic representation of the electron scattering, intensity distributions and final diffraction patterns from amorphous specimens of varying density and thickness. The shading indicates a higher density.

interact destructively, and no diffraction maxima will appear in the diffraction pattern from these scattered electrons. The resulting intensity distributions and the final diffraction patterns are shown schematically in Fig. 3.2, with the discrete intensity maxima, either diffraction spots or rings, superimposed on the diffuse background scattering.

The particular orientations at which sets of lattice planes with certain *d*-spacings will diffract the electron beam to form discrete diffraction spot or ring *reflections* are defined in the famous Bragg law,

$$\lambda = 2d_{hkl} \sin \theta,$$

where θ is the angle between the incident beam of electrons and the plane of atoms, as in Fig. 3.3, d_{hkl} is the spacing between the individual planes in this particular set of planes, and λ is the wavelength of the incident electrons. λ varies with the accelerating voltage applied to the electron gun, and at

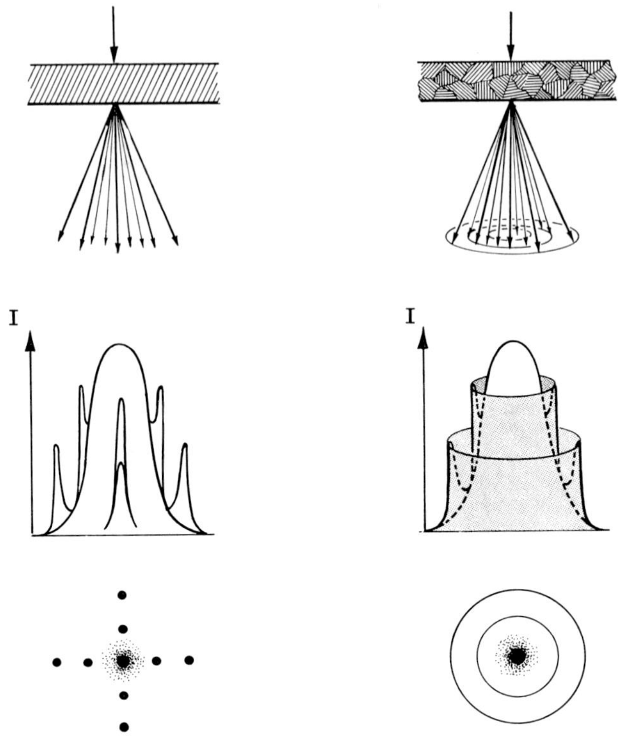

Fig. 3.2. Schematic representation of the electron scattering, intensity distributions and final diffraction patterns from a single crystal and a polycrystalline specimen.

80 kV, $\lambda = 0.004$ nm. A list of λ values for other accelerating voltages is given in Appendix 3.

Discrete diffraction spots or rings only occur when the incident electrons are diffracted by planes of atoms that lie at angles θ to the incident beam that satisfy the Bragg law. If the angle is such that the Bragg law is not satisfied, the scattered electrons will interact destructively, and only a small amount of diffuse scattering around the central spot in the diffraction pattern will appear. In practice, there are always some of the many sets of planes occuring in the crystal lying at the correct angles, called Bragg angles, and discrete reflections will always occur in the diffraction pattern of a crystalline specimen. The general term 'reflection' is often used to indicate either diffraction spots or rings, as they can be considered in simple terms to arise from the reflection of the electron beam from each set of lattice planes lying at a Bragg angle in the crystalline specimen (Figs. 4.1 and 4.6).

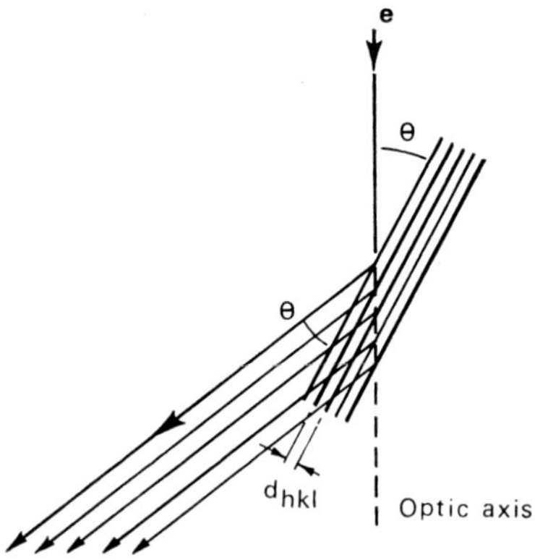

Fig. 3.3. Electron scattering at successive lattice planes of spacing d_{hkl}, lying at an angle θ to the incident electron beam.

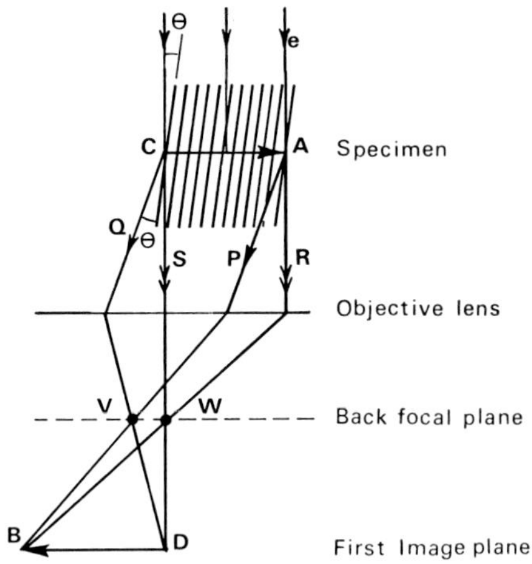

Fig. 3.4. Ray diagram tracing the paths of the diffracted and transmitted beams of electrons through the objective lens to the back focal plane and the first image.

Consider a single crystal specimen, with one set of planes lying at a Bragg angle, and follow the path of the electrons through the microscope, as in Fig. 3.4. It is axiomatic that whenever an image is formed in the image plane of the objective lens, a diffraction pattern occurs in the back focal plane. The formation of the diffraction pattern is shown in Fig. 3.4. In this diagram, rays leaving the same point of the specimen are brought to focus in the first image plane. Thus rays leaving point A in the specimen are brought to focus at B in the first magnified image, and rays leaving point C are focused at D. In addition, it can be seen that all rays scattered in the same direction by the specimen are brought to a focus in the back focal plane of the objective lens. So rays P and Q are brought to focus at point V, in the back focal plane, and rays R and S, which are not scattered, are brought to focus at point W. Thus rays that are scattered from the same *point* in the specimen are brought to a point focus in the *first image plane*, whereas rays that are scattered in the same *direction* from the specimen are brought to a point focus in the *back focal plane*. It is this 'image', formed in the back focal plane of the objective lens, that is known as the diffraction pattern.

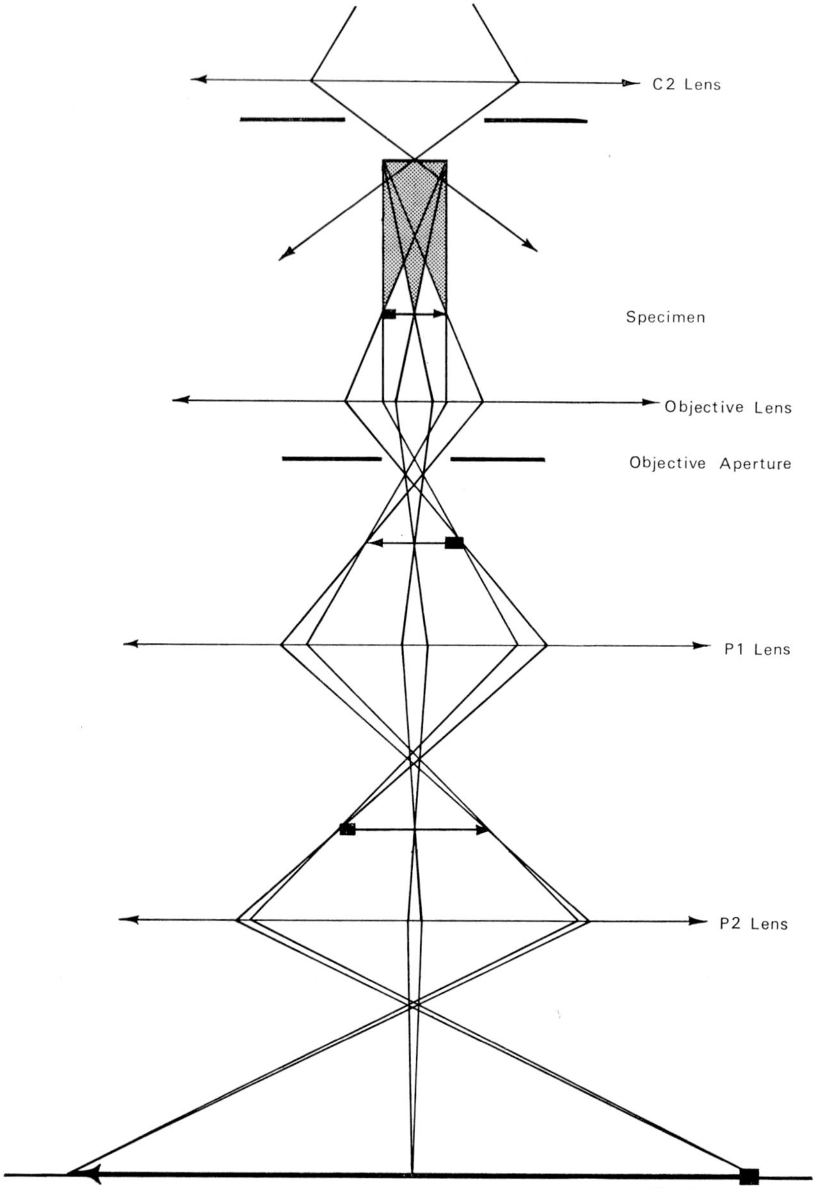

Fig. 3.5. Ray diagram showing the imaging mode of operation in an electron microscope having an objective and two projector lens imaging system.

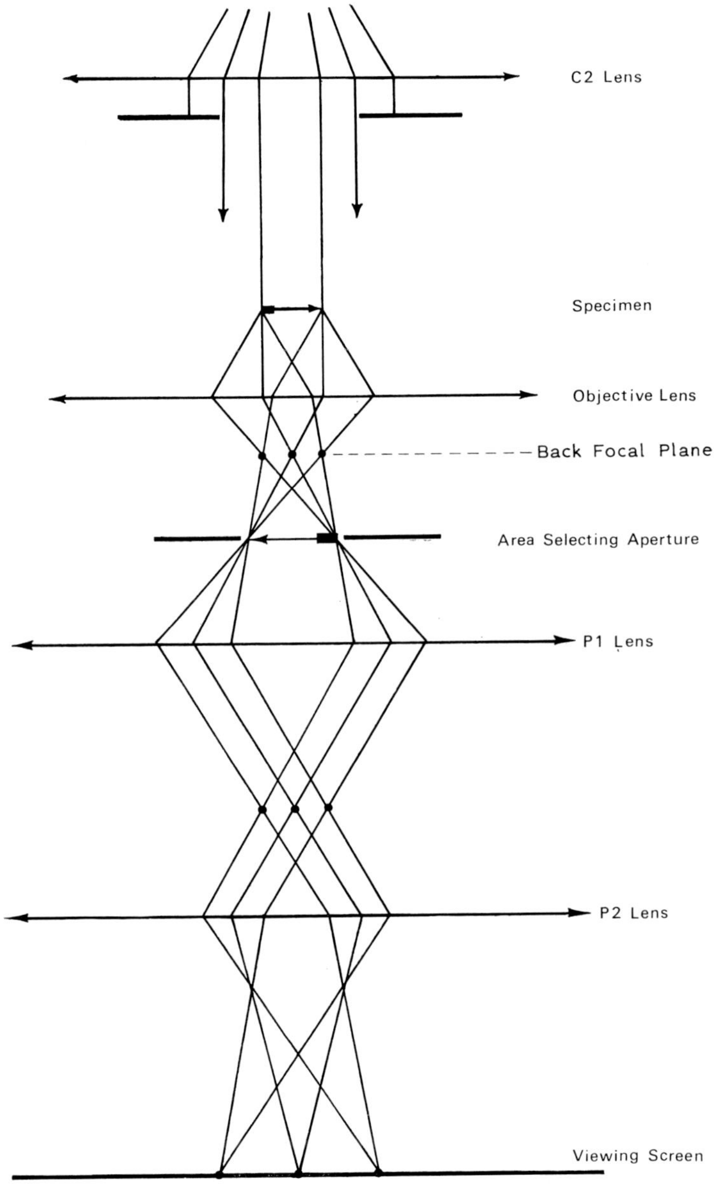

Fig. 3.6. Ray diagram showing the diffraction mode of operation in an electron microscope having an objective and two projector lens imaging system.

3.2 Observing the diffraction pattern

In the normal imaging or magnification mode of operation of a modern high resolution electron microscope the two or three projector lenses, which follow the objective lens, are focused to transfer the first image formed by the objective lens on to the viewing screen at a high magnification. For the sake of simplicify the description of diffraction pattern formation will be limited to the two projector lens system as the majority of electron microscopes presently operating have two projector lenses, and the three projector lens system is only a further extension of the principles of two projector lens operation. In the imaging mode of operation (Fig. 3.5), the first projector lens, P_1, is focused on the first image and forms a second image in the object plane of the second projector lens, P_2. P_2 then forms the final image on the viewing screen. In order to view the diffraction pattern, the P_1 lens must be weakened (operated with a lower current flowing through the lens coil) to increase the focal length so that it is focused on the back focal plane of the objective lens, rather than the first image plane. The diffraction pattern is then first focused in the object plane of P_2 and finally on the viewing screen (Fig. 3.6).

All modern microscopes have a convenient way of making the necessary adjustment of the P_1 lens for viewing the diffraction pattern, either automatically by pressing a button or turning a switch, or manually by operating a P_1 lens control until the diffraction pattern comes into focus on the viewing screen. With the automatic adjustment a fine control of P_1 lens current is also provided to allow fine focusing of the diffraction pattern. The necessary instructions for operating each electron microscope in the diffraction mode will be found in the microscope instruction manual. The instrument is often temporarily operated in this mode so that the objective aperture, which is inserted to enhance the contrast of the normal image, may be accurately centred around the central beam. This is necessary to avoid the introduction of gross astigmatism in the image which will be caused by misalignment of the objective aperture.

Having formed the diffraction pattern on the viewing screen, it may be necessary to make a permanent record on a photographic plate.

3.3 Recording the diffraction pattern

The diffraction pattern is recorded on a photographic plate or film in a similar manner to the normal image using identical emulsions, although a number of precautions should be observed. The electron beam is spread out over the whole area of the plate in the imaging mode, but is concentrated in the central spot of the diffraction pattern in the diffraction mode. The central spot is therefore very intense, and often very much more intense than the other diffraction spots or rings. In order that the central spot should not completely fog the central area of the plate during the photographic exposure, a beam stop is usually available to intercept the central beam. The beam stop should be placed in position before the exposure, and removed for a short period before the end of the exposure to produce an accurate record of the position of the central spot, as in Fig. 1.7.

Determination of the exposure required to record the diffraction pattern satisfactorily is often not as simple as for the normal image. Modern microscopes generally include a device for image exposure determination, but these are not always satisfactory for determining the necessary exposure for a diffraction pattern. The three common exposure meters are screen current detectors, photo-voltaic or photo-resistive cells and comparison spot photometers. The first two systems generally assume a fairly even image intensity over the whole screen or detecting area and calculate the exposure accordingly. The intensity in the diffraction pattern is obviously far from even over the screen, and such exposure meters may give false values.

A system which can detect the intensity over a small area is to be preferred, and is available in the small screens, or comparison spot photometers found on some instruments. However, because of the considerable variation in intensity between the central spot and the outer spots or rings, no detection system is particularly satisfactory, and the successful exposure of diffraction patterns is usually a matter of experimentation and experience. A good rule of thumb is to reduce the illuminating intensity, usually by defocusing the second condenser lens, C_2, until only the central spot is clearly visible on the screen and the other diffraction spots or rings have just about disappeared to the naked eye, and then to expose the plate for 30 sec.

The beam stop should be removed between five and ten seconds from the end of the exposure, depending upon the relative intensities of the central spot and the other diffraction spots in the diffraction pattern. This exposure technique has two particular advantages, firstly the exposure time is far less critical at these intensity levels, and secondly, the sharpness of the diffraction spots or rings is enhanced by the increase in the spatial coherence of the illumination on defocusing the C_2 lens. Defocusing the C_2 lens is the usual way of decreasing the illumination intensity in an electron microscope, but in some instruments it may be accomplished by decreasing the electron emission from the filament by reducing the filament voltage or increasing the gun bias. In this case the exposure time will again be less critical, but the sharpness of the diffraction pattern will not be improved, and it is therefore recommended that the C_2 lens, rather than the filament emission control, is used to achieve the correct exposure for recording a diffraction pattern.

3.4 Printing the diffraction pattern

Although accurate measurements of spacings between diffraction spots and rings should always be taken directly from the photographic plate, preferably using a microdensitometer or travelling microscope, there is often a need to produce positive prints and enlargements of diffraction patterns for reports and records.

The wide range of contrast present in a diffraction pattern can be recorded faithfully by the photographic plate, but presents considerable difficulty in reproduction in a print because of the much reduced tonal range in any single grade of photographic paper. Thus it is often difficult to reproduce diffraction phenomena close to and often masked by the scattering around the central spot, without losing some of the outlying spots. A technique for partially overcoming this difficulty has been described by Hamilton (1968). The technique, known as unsharp masking, consists of taking the diffraction pattern negative and printing a contact positive on to a second plate, and then aligning the two plates together in the photographic enlarger and making a print of the combined positive and negative. In this way the tonal range in the diffraction pattern is reduced to a level which can be accepted by a single grade of photographic paper. Fig. 3.7 shows the type of result that can be achieved by this technique, compared with the results of making two separate exposures, to print the inner rings at the expense of the outer ones, and vice-versa. The technique is particularly useful when the central spot is very bright and extensive, as when a diffraction beam stop has not been used; it is generally unnecessary when a beam stop has been used to reduce the recorded intensity of the central spot.

It must be remembered that photographic prints, particularly those produced by the full wet chemical process, often distort during preparation,

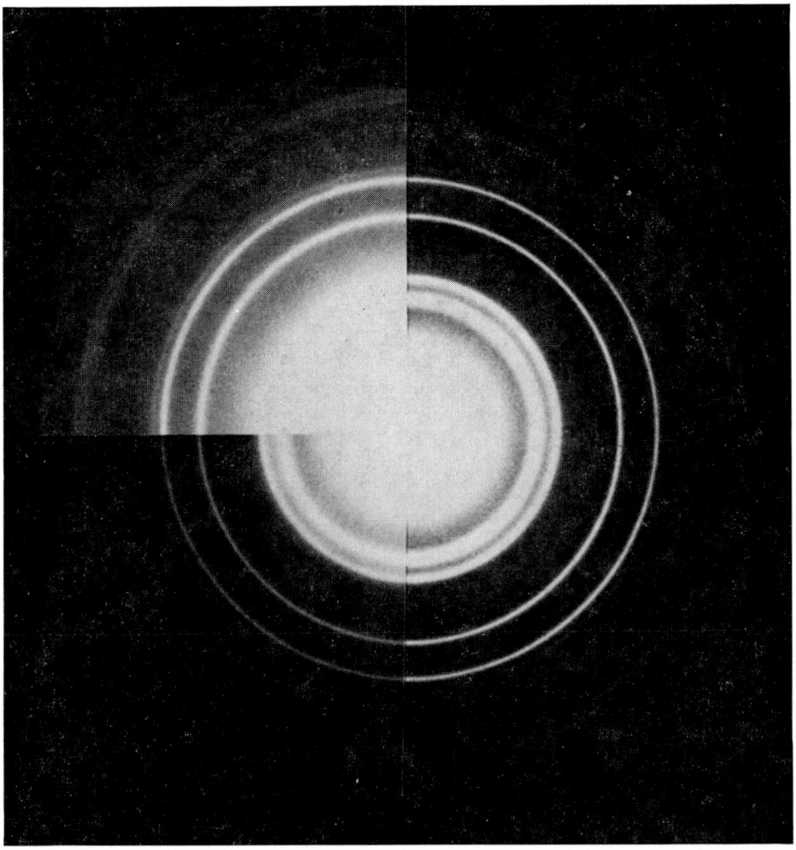

Fig. 3.7. A polycrystalline diffraction pattern showing, on the right, the type of result that can be achieved by unsharp masking, compared with the results of printing the negative to reveal preferentially either the inner or outer rings of the diffraction pattern, at the expense of the others. (From Hamilton 1968.)

particularly during glazing on a rotary drum. Prints prepared using stabilised paper and automatic processors (see List of suppliers) are usually rather less distorted. However, measurements upon which any reliance is to be placed should always be taken from the original photographic plate.

REFERENCE

Hamilton, J. F. (1968), Use of unsharp masking in printing electron micrographs, J. Appl. Phys. *39*, 5333.

FURTHER READING

Cohen, J. B. (1966), Diffraction methods in materials science (The Macmillan Comp., New York).

Rymer, T. B. (1970), Electron diffraction (Methuen and Co. Ltd., London).

Chapter 4

Analysis of the electron diffraction pattern

It is clear that a diffraction pattern contains two basic types of information, (a) spatial arrangment, that is, distances R, between the central spot and the other diffraction spots and rings, and the angles ϕ, between lines joining the central spot to each diffraction spot, and (b) intensities, in comparative values between the central spot and the other diffraction maxima.

There may, in addition, be other fine structural details in the diffraction pattern, but it is the distances, known as R values (generally measured directly off the photographic plate in millimeters), and the angles ϕ, which provide the readily interpretable information. From the R values it is possible to derive values of the d-spacings in the specimen, but before this can be done it is necessary to derive the simple formula connecting these two parameters.

4.1 Derivation of the basic formula, $Rd = \lambda L$

For the moment consider the electron microscope as a simple electron diffraction camera, with the electron beam striking a specimen and being diffracted to form a diffraction spot on the photographic plate, at distance R from the centre of the diffraction pattern, as in Fig. 4.1. For the time being, the effect of the presence of objective and projector lenses between the spemen and the plate is neglected; it will be consider later in § 5.3.1 .The distance between the specimen and the plate, the *camera length*, is designated L, and by simple geometry

$$\tan 2\theta = R/L.$$

Now the Bragg Law states that

$$\lambda = 2d \sin \theta,$$

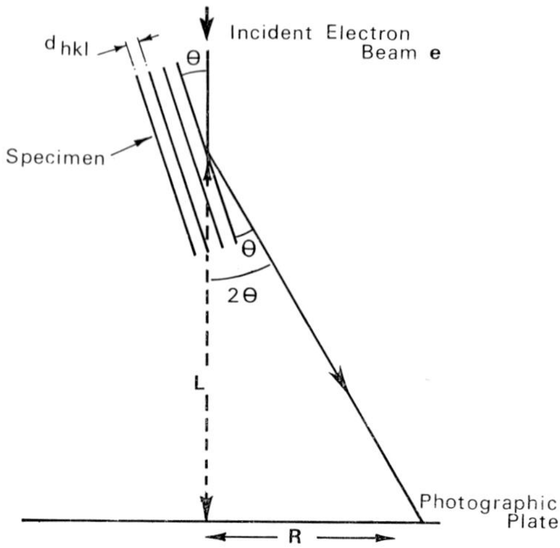

Fig. 4.1. The electron microscope considered as a simple electron diffraction camera.

and since the angles θ, through which the electrons are diffracted are very small, only 1°–2°, the approximation

$$\tan 2\theta = 2 \sin \theta,$$

can be made with very little error. Then

$$R/L = \lambda/d,$$

or

$$Rd = \lambda L.$$

So if values of R, L and λ for a particular diffraction spot or ring can be measured, then the d-spacing of the set of lattice planes giving rise to that spot or ring can be determined. This procedure can be repeated for all the spots or rings in the diffraction pattern, and a list of d-spacings can be compiled for all the sets of planes in the specimen which are reflecting. These d-spacings can then be used to determine a number of important pieces of information about the specimen.

4.2 Indexing a simple diffraction pattern from a known specimen

First the diffraction spots or rings must be indexed, that is, they must be assigned the Miller indices of the lattice planes from which they arise. Since a polycrystalline pattern is only the superposition of all the single crystal patterns from the specimen, the indexing procedure will be shown for single crystal patterns, and the procedure for polycrystalline patterns will then become apparent.

The procedure for indexing a single crystal diffraction pattern is dependent upon whether the composition and therefore the structure of the specimen is known, or whether the specimen is a completely unknown material. When the specimen structure is known an immediate choice of the correct formula correlating d, the lattice spacing, a, b, and c, the lattice parameters, α, β and γ, the angles between the major axes of the unit cell, and h, k and l, the Miller indices, can be made, from Appendix 2a. The procedure is to insert the known values for d, a, b and c, and α, β and γ, and thus determine the h, k and l values for each of the diffraction spots.

However, when the composition of the specimen, and therefore its structuere, are completely unknown the problem is very much more complex and no rigorous procedure can be laid down. Rather it is a matter of trial and error, combined with some intuition and whatever other limited knowledge of the specimen is available, the aim being to find the correct formula from Appendix 2a, which will give a consistent set of Miller indices, which when put back into the appropriate formula for the angles ϕ, from Appendix 2b, gives values which match the angles measured from the diffraction pattern.

For simplicity of explanation a diffraction pattern from a specimen known to have a cubic structure will be analysed. The procedure is very similar for other structures, with the substitution of the correct formula from Appendix 2a.

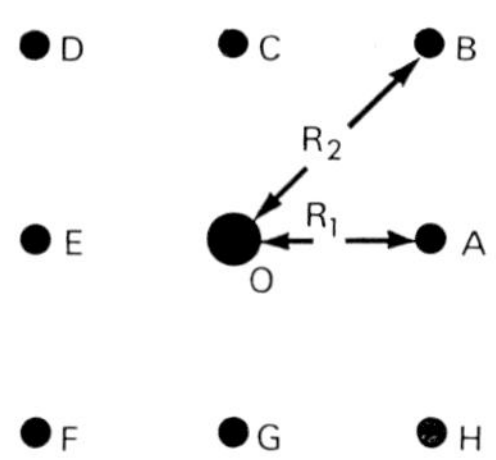

Fig. 4.2. Schematic diffraction pattern from a single crystal specimen with a simple cubic structure, of lattice parameter, $a = 0.25$ nm.

Fig. 4.2 is a schematic diffraction pattern from a single crystal specimen with a simple cubic structure of known lattice parameter, $a = 0.25$ nm. First the d-spacing for each diffraction spot must be determined, and then the Miller indices may be assigned.

4.2.1 DETERMINING THE *d*-SPACINGS

To begin, measure the distances between the central spot and each of the other spots in the diffraction pattern in turn. Since the pattern is symmetrical, this should be done by measuring between spots on either side of the central spot and dividing by two, or else by measuring the overall spacing of a large number of spots in a row passing through the central spot, and dividing by the number of spaces between the spots.

In the diffraction pattern in Fig. 4.2 it is apparent that A, C, E and G are all at the same distance from the centre as are B, D, F and H. It is therefore only necessary to measure the distances R_1 and R_2 marked on Fig. 4.2. In this pattern, $R_1 = 12.8$ mm and $R_2 = 18.1$ mm.

Now assume that the diffraction pattern was recorded with the microscope operating at an accelerating voltage of 80 kV, for which $\lambda = 0.004$ nm* (values of λ for other voltages are given in Appendix 3), and that the microscope was operating with a camera length, $L = 800$ mm.

So,

$$d_1\ (\text{nm}) = \frac{\lambda\ (\text{nm}) \times L\ (\text{mm})}{R_1\ (\text{mm})}$$

$$= \frac{0.004 \times 800}{12.8} = 0.25\ \text{nm},$$

and

$$d_2 = \frac{0.004 \times 800}{18.1} = 0.177\ \text{nm}.$$

4.2.2 ASSIGNING MILLER INDICES

In this example the specimen has a simple cubic structure, so using the formula for the cubic system from Appendix 2a,

$$\frac{1}{d^2} = \frac{h^2 + k^2 + l^2}{a^2}$$

and substituting $a = 0.25$ nm, the $(h^2 + k^2 + l^2)$ values of the two diffraction spots can be determined as follows.

* In practice λ may vary considerably, as apart from deliberate changes of the accelerating voltage to alter the image contrast perhaps, small fluctuations may also occur due to electronic instabilities. To eliminate inaccuracies caused by these small changes, λ and L are not determined separately, but grouped together as λL, the *camera constant*, and determined experimentally using known standards, as described in §§ 5.1 – 5.3.

Spot A

$$h_1^2 + k_1^2 + l_1^2 = \frac{a^2}{d_1^2} = \frac{(0.25)^2}{(0.25)^2} = 1$$

Spot B

$$h_2^2 + k_2^2 + l_2^2 = \frac{a^2}{d_2^2} = \frac{(0.25)^2}{(0.177)^2} = 2.$$

But h, k and l must be whole numbers, so

$h_1k_1l_1$ must be $\{100\}$, i.e. (100), (010) or (001).

and

$h_2k_2l_2$ must be $\{110\}$, i.e. (110), (101) or (011).

It is now possible to assign a set of Miller indices to the spots in the diffraction pattern. Proceed as follows;

(i) index the central spot as (000),

(ii) assign, quite arbitrarily, since they are all equivalent, one of the three possibilities for $(h_1k_1l_1)$ as the index for spot A, say (100),

(iii) assign one of the three possibilities for $(h_2k_2l_2)$ to index spot B. Here there are some restrictions. The indices of all the spots in the diffraction pattern must be consistent with vectorial addition of the individual indices. That is, one may think of getting to spot B in the diffraction pattern by taking one step in the direction OA and then one step in the direction OC, as in Fig. 4.3. In this example, taking a step in the direction OA, from the central spot (000) to spot A now indexed (100), adds 1 to the value of the h index. So, spot B, which also involves one step in the OA direction must also have its h index equal to 1. Therefore spot B may be indexed either (110) or (101), but not (011). So, arbitrarily assign (110) to index spot B, as in Fig. 4.4.

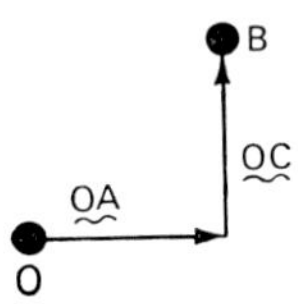

Fig. 4.3. Vectorial addition of Miller indices.

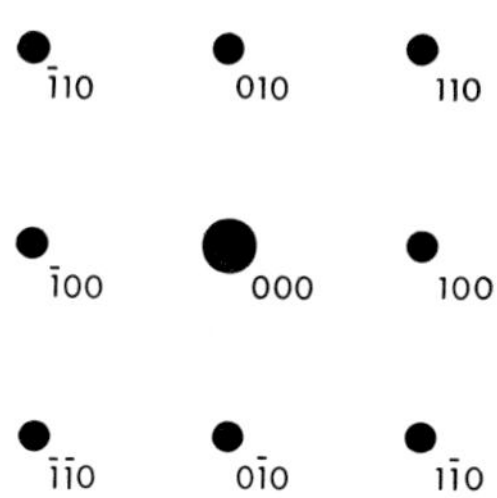

Fig. 4.4. A vectorially consistent set of Miller indices for the simple cubic diffraction pattern in Fig. 4.2.

Having assigned indices to two spots in the diffraction pattern, other than the central spot, the indices for all the other spots are automatically determined. Assigning k_2 as 1 dictates that a step in the OC direction adds 1 to the k index, and conversely a step in the CO or OG direction deducts 1 from the k index. So, spot C is indexed (010) and spot G is indexed $(0\bar{1}0)$, where a bar over the number indicates a minus value. It is now possible to complete the indexing of the diffraction pattern, as in Fig. 4.4.

Two points about the assignment of these indices should be noted,

(a) in going along any straight line of spots, if a change in the h, k or l index occurs in one step, a similar change must also occur on the next step, e.g. in going from $(\bar{1}10)$ to $(\bar{1}00)$, no change in the h and l indices occurs, but 1 is deducted from the k index. So on the next step 1 must also be deducted from the k index, giving, as in Fig. 4.4, $(\bar{1}\bar{1}0)$,

(b) spots equidistant either side of the central spot, along a straight line, are the negative of each other, e.g. $(\bar{1}\bar{1}0)$ is the negative of (110). Using these two rules and the vectorial addition explained previously, that is, for example

$$(010) + (100) \rightarrow (110),$$

it is very easy to index the complete pattern, once indices have been assigned to the first two spots.

4.2.3 CHECKING THE INTERPLANAR ANGLES

For completeness, a check should be made that the angles between a number of the spots in the diffraction pattern are correct for the indices that have been assigned. From Appendix 2b the angle ϕ between the two lines joining two diffraction spots to the central spot, in the cubic system, is given by

$$\cos\phi = \frac{h_1h_2 + k_1k_2 + l_1l_2}{\sqrt{[(h_1^2 + k_1^2 + l_1^2)(h_2^2 + k_2^2 + l_2^2)]}}$$

Now, for the diffraction spots A and C in Fig. 4.2, for example, assigned (100) and (010), $\cos\phi$ is

$$\cos\phi_{(100)\,\widehat{}\,(010)} = \frac{1\times 0 + 0\times 1 + 0\times 0}{\sqrt{[(1+0+0)(0+1+0)]}} = 0$$

and so

$$\phi_{(100)\,\widehat{}\,(010)} = 90^\circ$$

as indeed it is in the diffraction pattern as indexed in Fig. 4.4. Also

$$\cos\phi_{(100)\,\widehat{}\,(110)} = \frac{1\times 1 + 0\times 1 + 0\times 0}{\sqrt{[(1+0+0)(1+1+0)]}} = \frac{2}{\sqrt{2}}$$

and so

$$\phi_{(100)\,\widehat{}\,(110)} = 45^\circ$$

as again it is in Fig. 4.4.

For a simple diffraction pattern with low order reflections, that is diffraction spots whose Miller indices contain only small numbers, (e.g. 100, 110, 210 ...), this check of the angles is generally unnecessary, though it can be very important for some higher order reflections and more complex diffraction patterns. If the procedure in §§ 4.2.1 and 4.2.2 gives values of $(h^2 + k^2 + l^2)$ equal to 9, 18, 27 ... there are two possible sets of indices that may be assigned to these reflections.

$(h^2 + k^2 + l^2) = 9$ may be indexed either (221) or (300)

$(h^2 + k^2 + l^2) = 18$ may be indexed either (411) or (330)

and

$(h^2 + k^2 + l^2) = 27$ may be indexed either (511) or (333)

as in Table 4.1. It may be possible to decide which of the two possible sets of indices is correct in each case, simply by inspection of the indices of the neighbouring diffraction spots, but the assigned indices should always be checked by substituting the figures in the appropriate formula for the interplanar angles from Appendix 2b.

4.2.4 ALTERNATIVE PROCEDURE FOR CUBIC CRYSTALS – THE RATIO METHOD

For cubic crystals a simple ratio method may be used for determining the indices of diffraction spots directly from the R values, without the need for determination of the camera constant λL.

Since

$$Rd = \lambda L$$

and for cubic crystals

$$\frac{1}{d^2} = \frac{h^2 + k^2 + l^2}{a^2}$$

$$\frac{1}{d^2} = \left(\frac{R}{\lambda L}\right)^2 = \frac{h^2 + k^2 + l^2}{a^2}$$

and

$$R = \sqrt{(h^2 + k^2 + l^2)} \times \lambda L/a$$

TABLE 4.1

Possible $(h^2 + k^2 + l^2)$ values

$(h^2 + k^2 + l^2) = N$	$\{h, k, l\}$
1	100
2	110
3	111
4	200
5	210
6	211
7	–
8	220
9	221 or 300
10	310
11	311
12	222
13	320
14	321
15	–
16	400
17	410
18	411 or 330
19	331
20	420
21	421
22	332
23	–
24	422
25	500
26	510
27	511 or 333
...	...

Then in the ratio R_1/R_2, the constant factor $\lambda L/a$ cancels out, and

$$\frac{R_1}{R_2} = \frac{\sqrt{(h_1^2 + k_1^2 + l_1^2)}}{\sqrt{(h_2^2 + k_2^2 + l_2^2)}}$$

Since h, k and l must all be whole numbers, $(h^2 + k^2 + l^2)$ must be equal to one of the integers in Table 4.1. In fact, $(h^2 + k^2 + l^2)$ may be any whole number except $7n + (n - 1)$, where n is an integer, for clearly 7, 15, 23 ... cannot be made up of the sum of the squares of three whole numbers.

So the ratio of the R values must equal the ratio of the square roots of two whole numbers. A simple way of determining the two whole numbers $(h_1^2 + k_1^2 + l_1^2)$ and $(h_2^2 + k_2^2 + l_2^2)$ is to determine the ratio of the two R values, R_1/R_2, by division on the C and D scales of a standard slide-rule, and then find the two whole numbers which are opposite each other on the A and

B scales. These two numbers will be $(h_1^2 + k_1^2 + l_1^2)$ and $(h_2^2 + k_2^2 + l_2^2)$, and it is a simple matter to read off the appropriate values of $\{h, k, l\}$ from Table 4.1.

Further spots in a cubic diffraction pattern are then rapidly indexed by measuring their R values, and adopting the same procedure, possibly using one of the previously determined $(h^2 + k^2 + l^2)$ values as a reference point each time. If frequent indexing of cubic diffraction patterns is to be undertaken, it may be considered worthwhile making a special slide-rule with (h, k, l) indices against the appropriate whole numbers on the A and B scales of the slide-rule (Roser and Thomas 1964).

4.3 Determining the zone axis of the diffraction pattern

In the simple diffraction pattern indexed in § 4.2.2, it is noticable that the l index for all the diffraction spots is 0. Referring back to Fig. 2.8, it is clear that if l is 0 then the planes do not intersect the z axis, or in other words, all the planes are parallel to the z axis. So these are the planes that are giving rise to the diffraction spots in this diffraction pattern. If the electron beam is to be diffracted from all of these planes, as in Fig. 4.5, it must strike them all at a very shallow angle, less than 1° or 2°. The only way that this can happen is if the beam is 'travelling down' the z axis. Considering each of the planes individually, as in Fig. 4.6, with the electron beam diffracted from both the 'front' and 'back' of each of the planes, the origin of each of the eight diffraction spots in Fig. 4.4 is quite clear. The centre spot is formed by those electrons which are not scattered and pass straight through the crystal. Thus it is established that the specimen giving rise to this diffraction pattern is oriented with the electron beam parallel to the z axis, that is,

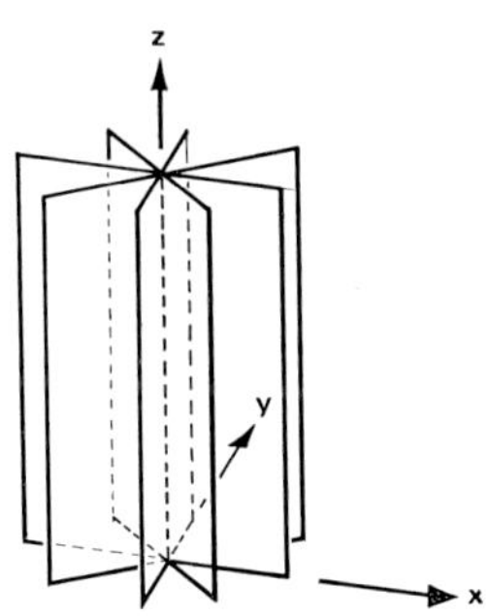

Fig. 4.5. Planes lying parallel to the z axis in the crystal lattice.

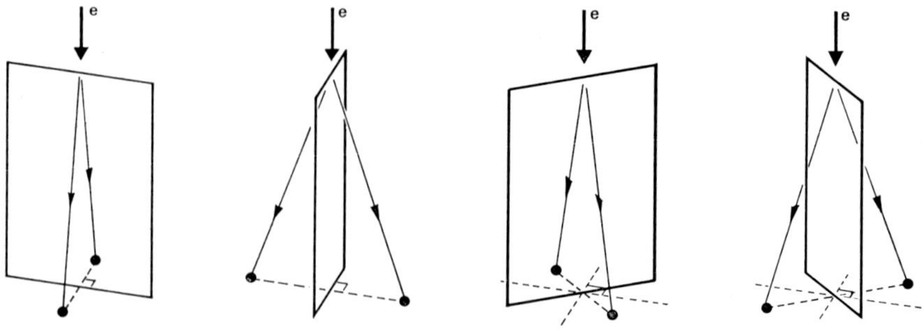

Fig. 4.6. Individual planes parallel to the z axis, diffracting the incident electron beam.

perpendicular to the (001) planes. So the orientation of the specimen is defined by stating that the zone axis of the diffraction pattern is [001], where [ ] denotes a particular direction in the crystal.

In addition to this qualitative derivation, a rapid mathematical derivation of the zone axis $[U, V, W]$ is possible, using the following expressions

$$U = k_1 l_2 - k_2 l_1$$
$$V = l_1 h_2 - l_2 h_1$$
$$W = h_1 k_2 - h_2 k_1,$$

where $h_1 k_1 l_1$ and $h_2 k_2 l_2$ are the indices of any two of the spots in the diffraction pattern, other than the central spot or two spots either side of the central spot. A simple way of doing the mathematics is to write down the indices twice, one set above the other, strike out the first two and the last two figures,

$$\begin{array}{c|ccccc|c} h_1 & k_1 & & l_1 & & h_1 & & k_1 & l_1 \\ & & \times & & \times & & \times & & \\ h_2 & k_2 & & l_2 & & h_2 & & k_2 & l_2 \end{array}$$

and then cross-multiply in pairs, deducting the second multiple from the first, for the three groups of four, from left to right, as follows.

$$k_1 l_2 - l_1 k_2, \qquad l_1 h_2 - h_1 l_2, \qquad h_1 k_2 - k_1 h_2.$$

If this is done for any two spots in the diffraction pattern in Fig. 4.4, for example $(0\bar{1}0)$ and $(1\bar{1}0)$, the zone axis is given by

$$\begin{array}{r|ccccccc|l} 0 & \bar{1} & \times & 0 & \times & 0 & \times & \bar{1} & 0 \\ 1 & \bar{1} & & 0 & & 1 & & \bar{1} & 0 \end{array}$$

$$(0 - 0), \qquad (0 - 0), \qquad (0 - \bar{1}),$$

that is [001].

For other combinations of indices the zone axis might have been calculated to be $[00\bar{1}]$, but this is equivalent to [001].

4.4 Indexing a diffraction pattern with many diffraction spots

Fig. 4.7 is a schematic diffraction pattern from a simple cubic single crystal, identical to Fig. 4.2, in exactly the same orientation, but now with many more diffraction spots appearing. The inner diffraction spots are immediately indexed from Fig. 4.4, and then by following the same procedure as previously the remaining spots are easily indexed.

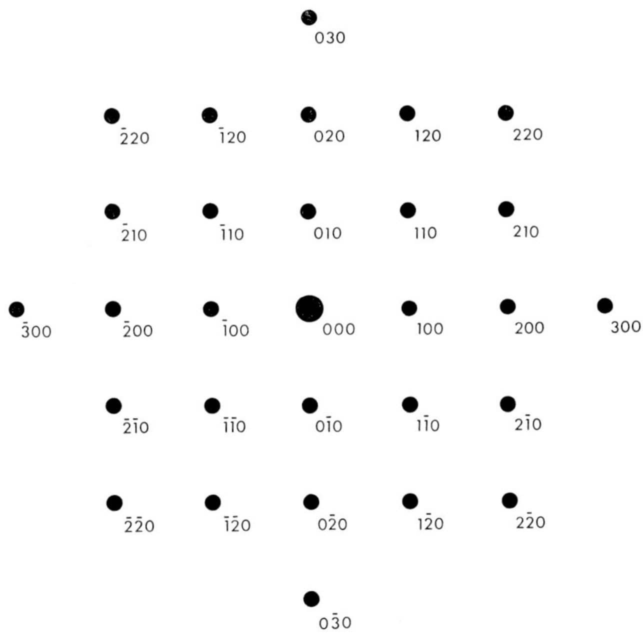

Fig. 4.7. Schematic diffraction pattern from a simple cubic single crystal specimen, with many diffraction spots, indexed as Fig. 4.4.

The conditions were that, having assigned A as (100) and B as (110),
(i) a step in the OA direction adds 1 to the h index,
(ii) a step in the OC direction adds 1 to the k index, and
(iii) the l index is always zero, in this particular diffraction pattern.
The complete diffraction pattern is now indexed, as in Fig. 4.7.

4.5 Forbidden diffraction spots

In the diffraction patterns from certain crystal lattices, some of the expected diffraction spots may not occur, and are termed forbidden reflections. The diffraction pattern from the simple lattice has no forbidden reflections, and the full diffraction pattern occurs. However, if for example, the corresponding diffraction pattern, with [001] zone axis, is formed for a face-centred cubic specimen, it will be seen that many reflections found in Fig. 4.4 are missing. Fig. 4.8 is a schematic [001] diffraction pattern from a pure nickel

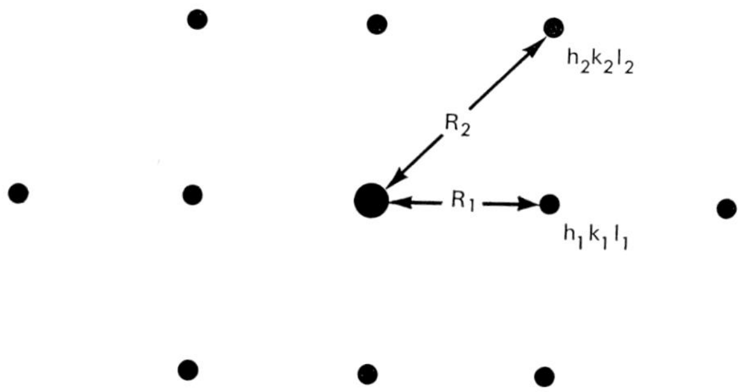

Fig. 4.8. Schematic diffraction pattern from a single crystal nickel specimen, $a = 0.35$ nm.

specimen, $a_{Ni} = 0.35$ nm. Measurement of the R values and computation using values of $\lambda = 0.004$ nm and $L = 800$ mm as before, gives

$$d_{h_1k_1l_1} = 0.176 = \frac{a}{\sqrt{(h_1^2 + k_1^2 + l_1^2)}}$$

$$d_{h_2k_2l_2} = 0.124 = \frac{a}{\sqrt{(h_2^2 + k_2^2 + l_2^2)}}$$

such that

$$(h_1^2 + k_1^2 + l_1^2) = 4$$

and

$$(h_2^2 + k_2^2 + l_2^2) = 8.$$

Now the only indices which will fit these requirements are

$$h_1k_1l_1 \text{ of } \{200\} \text{ type}$$
$$h_2k_2l_2 \text{ of } \{220\} \text{ type}$$

and so the diffraction pattern is indexed as in Fig. 4.9.

Alternate diffraction spots are now absent from the diffraction pattern and are termed forbidden reflections. The basis for certain reflections being forbidden is made clear in Fig. 4.10. In this lattice it is possible to draw an extra plane PQRS, containing the face-centering atoms, in which the number and the arrangement of the atoms is precisely the same as in the original (100)

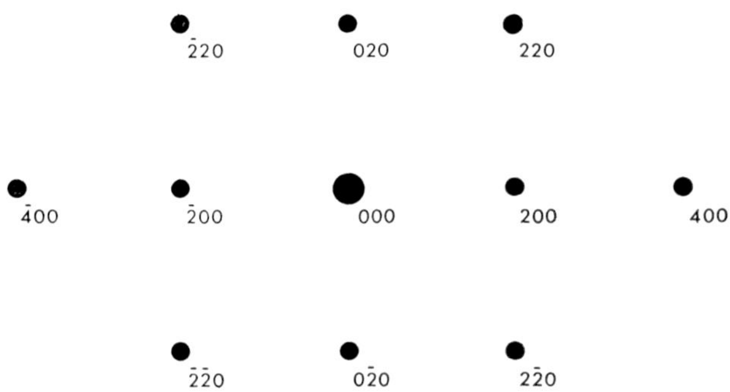

Fig. 4.9. Consistent set of Miller indices for the diffraction pattern in Fig. 4.8.

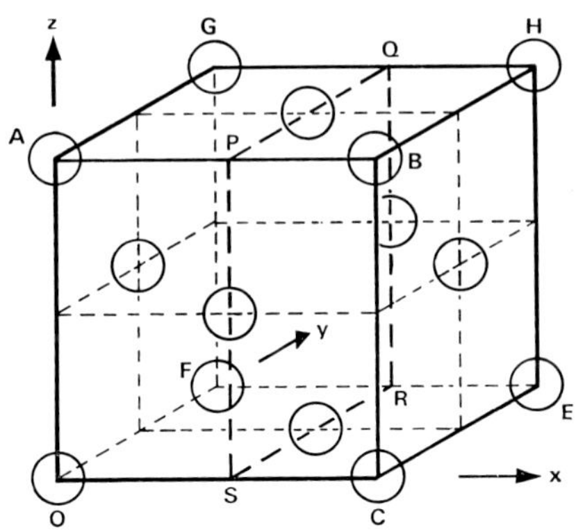

Fig. 4.10. Unit cell of the face-centered cubic lattice.

planes OAGF and CBHE, and is exactly half way between them. This plane is indexed (200), as it cuts the x, y, and z axes at $a/2$, ∞ and ∞ respectively, as described in § 2.2.4. Now of course the electron beam cannot distinguish between planes made up of the corner atoms of the unit cell and those made up of the face-centering atoms, as the two types of planes are identical, and the point chosen for the origin of the unit cell is purely arbitrary. So now there are twice the number of planes, indexed as (200) planes at $a/2$ lattice spacing, and (100) planes no longer exists as separate entities. Similarly (020) and (002) planes can be drawn in the lattice and (010) and (001) planes no longer occur.

As the electron beam no longer encounters atomic planes of the type {100} with lattice spacings, a, diffraction spots corresponding to these planes no longer appear in the diffraction pattern and are effectively forbidden. Selection rules for the cubic system allow reflections to appear in the diffraction pattern only when

for the f.c.c. lattice h, k and l are ALL EVEN
and for the b.c.c. lattice $(h + k + l)$ is EVEN

4.6 Superlattice reflections

Consider now a specimen containing two different elements – a binary alloy – in which the atomic sites of the unit cell are shared between the atoms of the two elements. If the two elements are distributed randomly throughout the specimen, the atomic sites in the unit cell will be occupied by atoms of either element entirely at random, in proportion to the percentage elemental composition of the specimen. The diffraction pattern from such an alloy will be very similar to that from a single element specimen with the same structure, except that the R values and the intensities of the diffraction spots may be somewhat different.

Frequently the different elements of such an alloy are not randomly distributed throughout the specimen however, and the atoms of one of the elements may occupy one particular site in the unit cell, while the atoms of the other element occupy another discrete site. The specimen is then said to be ordered and the structure is termed a superlattice, e.g. an f.c.c. superlattice or a b.c.c. superlattice. Ordering generally occurs when the fractional amounts of the elements constituting the alloy are simple numbers, 1:1, 2:1, 3:1 .... Examples of ordered alloys in which superlattices occur are a copper-gold alloy, Cu_3Au, an f.c.c. superlattice, and cesium chloride, CsCl, a b.c.c. superlattice, as shown in Fig. 4.11.

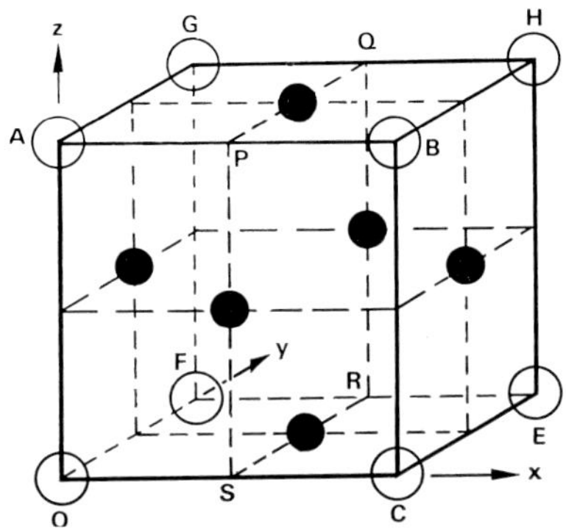

Fig. 4.11a. f.c.c. superlattice, Cu_3Au.

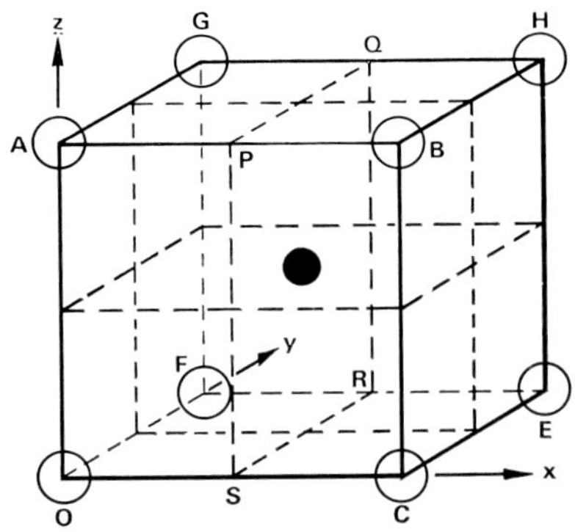

Fig. 4.11b. b.c.c. superlattice, CsCl.

In the f.c.c. superlattice, the copper atoms occupy the face-centred sites (each of the face-centred sites is shared equally between two neighbouring unit cells, so only half of each site is in one particular unit cell, i.e. $6 \times \frac{1}{2} = 3$, so Cu_3), and the gold atoms occupy the corner sites of the unit cell (each of the corner sites is shared equally between eight neighbouring unit cells, so only $\frac{1}{8}$ of each site is in this unit cell, i.e. $8 \times \frac{1}{8} = 1$, so Cu_3Au).

Now in the f.c.c. superlattice, the {200} planes are alternately made up of an equal number of copper and gold atoms ($1 \times \frac{1}{2} = \frac{1}{2}$ an atom of copper, and $4 \times \frac{1}{8} = \frac{1}{2}$ an atom of gold) as in OAGF and CBHE, and solely of copper atoms, as in SPQR. The scattering efficiency of these two different sets of planes will not be the same, and the electron beam will see two distinguishable types of plane, one with a spacing a from the origin, that is {100}, and one with a spacing $a/2$ from the origin that is {200}. The [001] diffraction pattern will then be basically that of an f.c.c. structure, as Fig. 4.9, since that is the structural arrangement of the atoms, but in addition, the intermediate spots of the simple cubic pattern will appear, with intensities determined by the difference in the structure factors of the two elements F_{Cu}–F_{Au}, as in Fig. 4.12, which is a schematic [001] diffraction pattern of a Cu_3Au specimen. Thus the forbidden reflections may reappear, if the composition of the planes in the unit cell varies in an orderly manner, even though they would not be expected from the structural arrangement of the atom sites.

Similarly, in the b.c.c. superlattice, CsCl, the {200} planes alternately consist solely of cesium and solely of chlorine atoms, and so {100} reflections will appear in the [001] diffraction pattern, in addition to the basic {200} reflections of the normal b.c.c. structure, as shown schematically in Fig. 4.13.

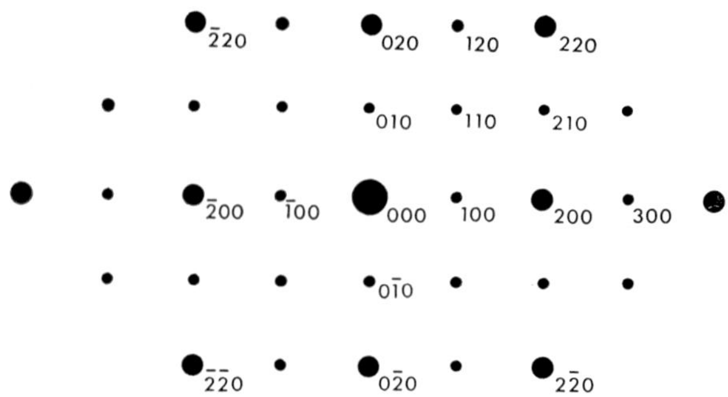

Fig. 4.12. Schematic [001] diffraction pattern from the f.c.c. superlattice specimen, Cu_3Au, in Fig. 4.11a.

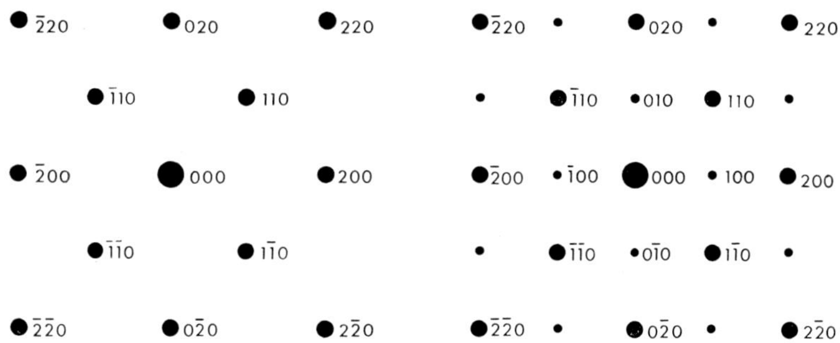

Fig. 4.13. Schematic [001] diffraction patterns from the basic b.c.c. structure and the b.c.c. superlattice, CsCl.

Diffraction patterns with zone axes other than [001] may also contain superlattice reflections depending on whether or not the alternate lattice planes can be discriminated due to the presence of different atoms in the reflecting planes. For example, the diffraction patterns from the same f.c.c. and b.c.c. superlattices, but now with [110] zone axes, that is with the crystals oriented in the electron microscope so that the electron beam is incident on one of the edges of the unit cell as in Fig. 4.14a, also contain extra reflections as shown in Fig. 4.14b.

In the f.c.c. superlattice, the electron beam can distinguish between the ($\bar{2}20$) planes, SPUT and VWQR, and the ($\bar{1}10$) plane OAHE, as the former contain only copper atoms while the latter contains a mixture of both copper and gold atoms, and so diffraction spots from both types of plane occur in the diffraction pattern. Also, the (001) and (002) are distinguishable, as the former contains both types of atoms while the latter contains only copper

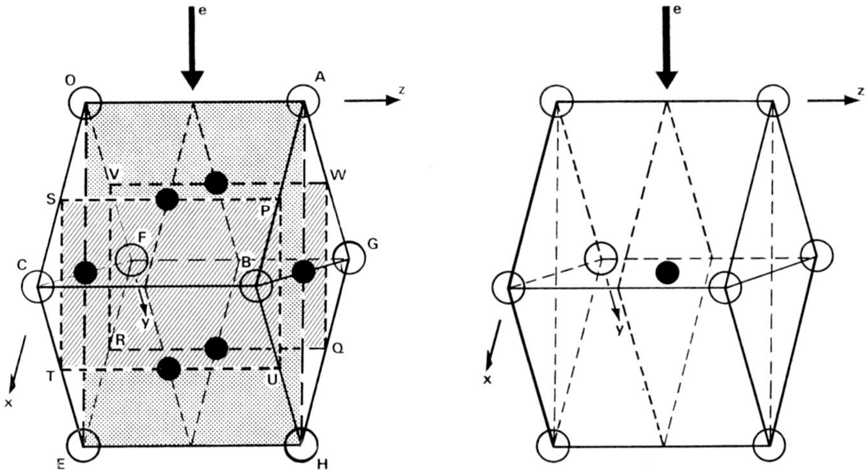

Fig. 4.14a. The f.c.c. superlattice, Cu_3Au, and b.c.c. superlattice, CsCl, oriented with (001) and (002) planes parallel to the incident electron beam.

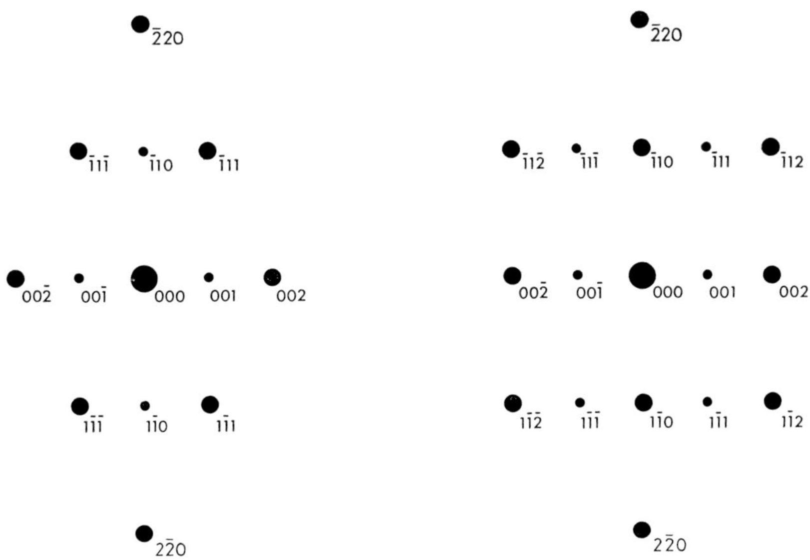

Fig. 4.14b. Schematic [110] diffraction patterns from the f.c.c. superlattice, Cu_3Au, and the b.c.c. superlattice, CsCl.

atoms, so spots will appear in the diffraction pattern for both types of lattice planes.

Similarly, the (001) and (002) planes in the b.c.c. superlattice are distinguishable in this orientation and are represented separately in the diffraction pattern.

4.7 Reciprocal lattice and Ewald sphere construction

The simple diffraction patterns so far considered have been analysed using a combination of simple geometry and three-dimensional observation of the crystal unit cell. Now interpretation in this manner is often slow and becomes very difficult for more complex crystal types, and at orientations away from the major crystal axes. It is preferable then to make use of two geometrical concepts, which though strange at first, once mastered, make analysis of many diffraction patterns possible at a glance. First it is necessary to construct a lattice of points, each one of which represents a set of planes in the real crystalline lattice, and then to apply a geometric construction to discover which of these sets of planes is reflecting and will therefore give rise to a diffraction spot in the diffraction pattern.

4.7.1 THE RECIPROCAL LATTICE

The reciprocal lattice is directly related to the real lattice, and is constructed by drawing a normal to each set of planes in the real crystal lattice ,and marking off points along these normals at distances $1/d$, from the origin, where d is the interplanar spacing for each set of planes. Thus the (200) planes of the f.c.c. lattice give rise to a point ●200, in the reciprocal lattice

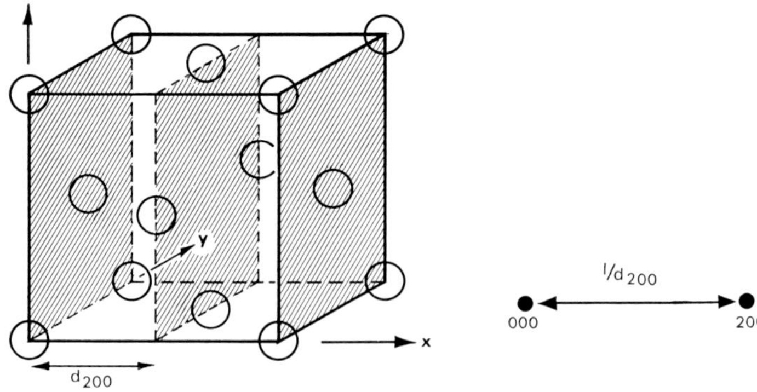

Fig. 4.15a. (200) Planes in the f.c.c. unit cell and the corresponding reciprocal lattice point.

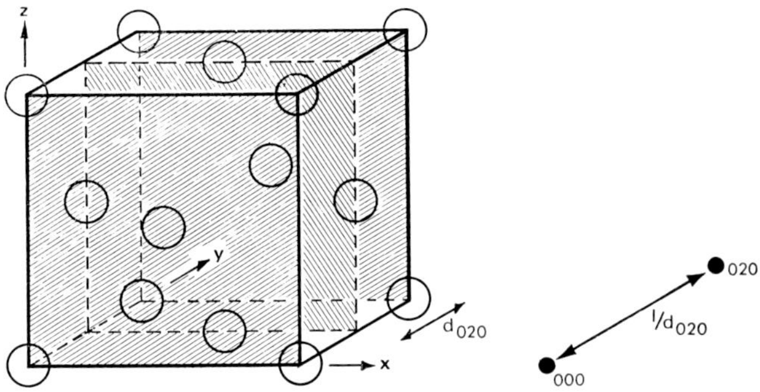

Fig. 4.15b. (020) Planes in the f.c.c. unit cell and the corresponding reciprocal lattice point.

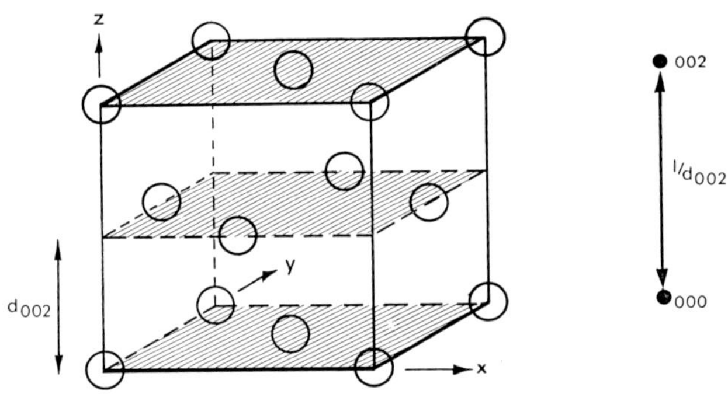

Fig. 4.15c. (002) Planes in the f.c.c. unit cell and the corresponding reciprocal lattice point.

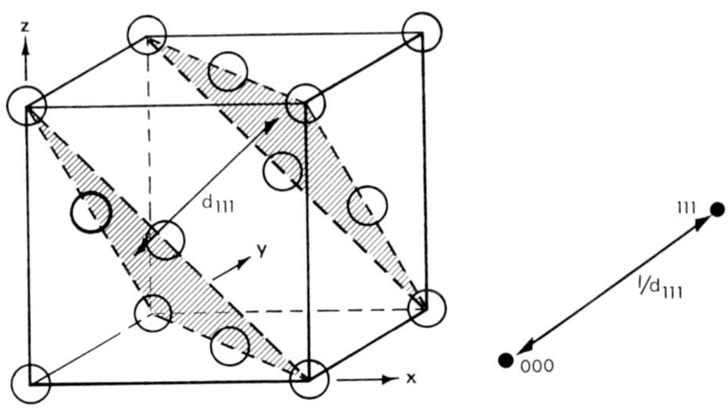

Fig. 4.15d. (111) Planes in the f.c.c. unit cell and the corresponding reciprocal lattice point.

at a distance $1/d_{200}$, from the origin ●000, in a direction normal to the (200) planes, as in Fig. 4.15a. Similarly the (020) planes, Fig. 4.15b, the (002) planes, Fig. 4.15c and the (111) planes, Fig. 4.15d, give rise to ●020, ●002 and ●111 points in the reciprocal lattice. When these points are combined they give the basic array of the f.c.c. reciprocal lattice (Fig. 4.16). This basic array can

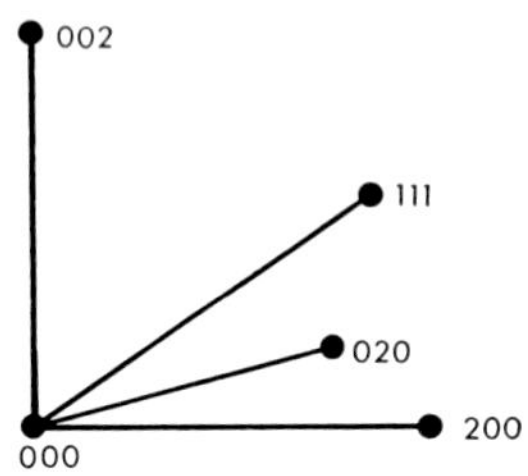

Fig. 4.16. The basic array of the f.c.c. reciprocal lattice.

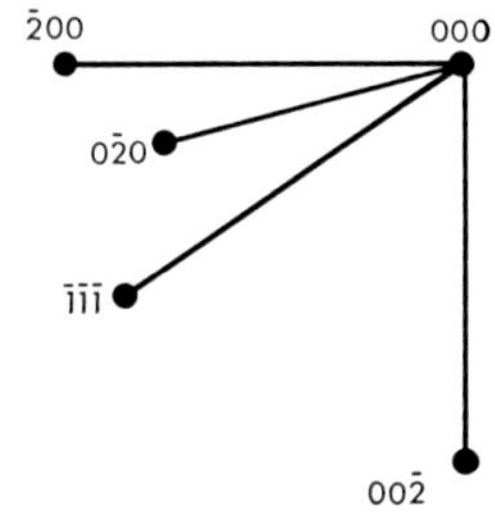

Fig. 4.17. Reciprocal lattice points corresponding to the reverse sides of the planes represented by the points in Fig. 4.16.

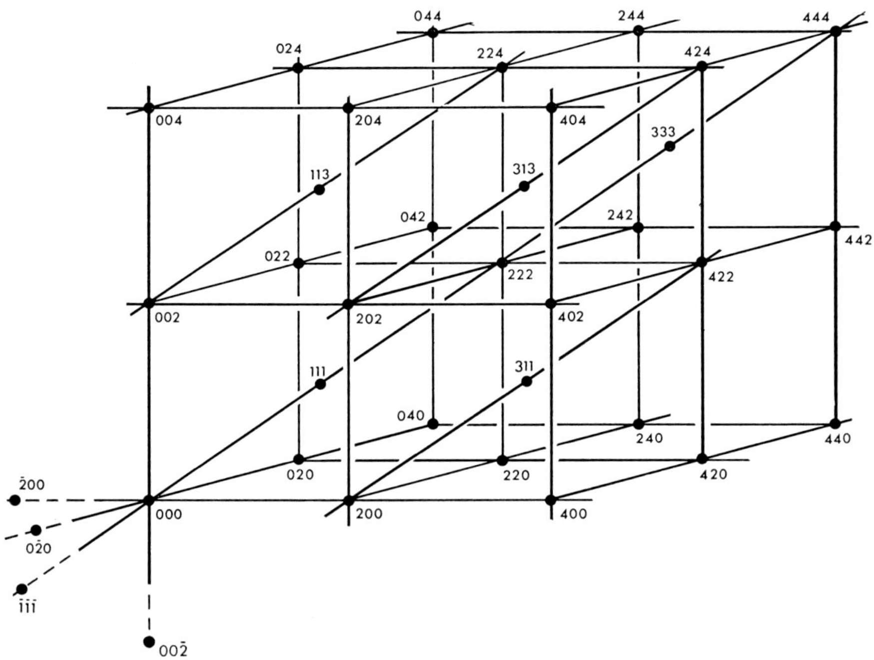

Fig. 4.18. A sector of the full f.c.c. reciprocal lattice. The other sectors can be imagined extending in the directions of the dotted lines.

be extended indefinitely by the addition of points for all the other sets of planes which it is possible to draw in the f.c.c. real lattice. Further points are added to represent the reverse sides of the planes to those already considered, as in Fig. 4.17, where the indices include a bar over the digits to indicate that they are positioned on the negative side of the origin in the reciprocal lattice. So the full reciprocal lattice for the f.c.c. crystal lattice can now be drawn (Fig. 4.18), and is seen to be identical to a real b.c.c. lattice, with the atom positions substituted by reciprocal lattice points. In fact the f.c.c. reciprocal lattice is a b.c.c. array of points and the b.c.c. reciprocal lattice is an f.c.c. array of points.

The full three-dimensional reciprocal lattice appears complex at first, but a section through the lattice, that is, a two-dimensional slice out of the lattice,

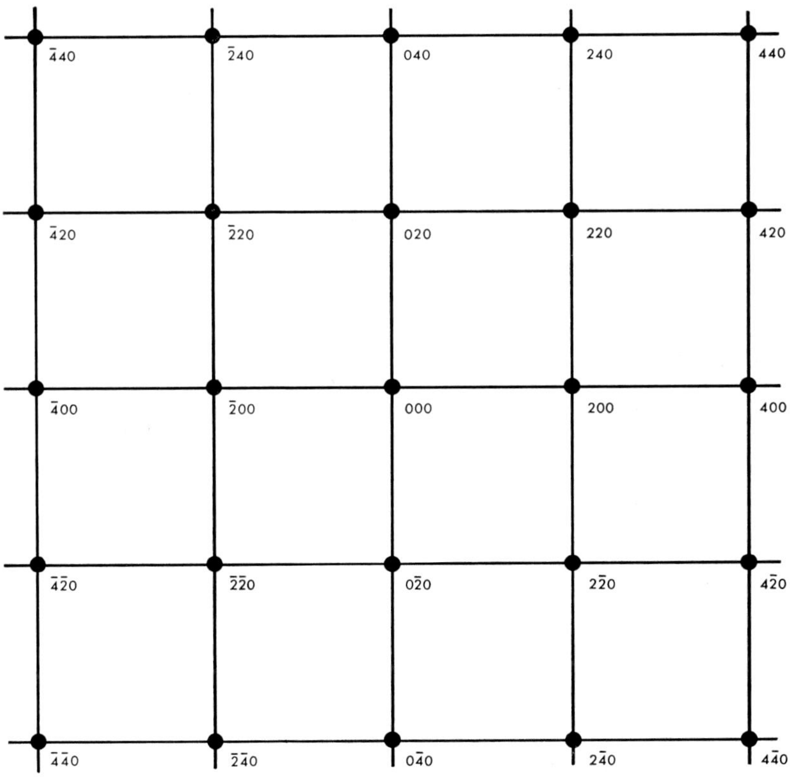

Fig. 4.19. [002] Reciprocal lattice section taken from the full reciprocal lattice shown in Fig. 4.18.

is easily drawn on a plane sheet of paper, and is much more quickly appreciated. Fig. 4.19 is such a reciprocal lattice section taken from Fig. 4.18, by selecting the plane of points through the origin, normal to [002]. It is now readily observed that this array of reciprocal lattice points is identical to the array of diffraction spots in the diffraction pattern in Fig. 4.9. Other sections through the reciprocal lattice are now easily drawn, particularly if a ball-and-rod model of the full reciprocal lattice is made, with the individual balls marked with the correct Miller indices. Any diffraction pattern for an f.c.c. specimen can now be indexed by choosing the reciprocal lattice section with an identical spatial arrangement of points to that of the spots in the diffraction pattern.

Reciprocal lattice models may be made for the other possible crystal structures which are likely to be confronted in the analysis of diffraction patterns, and two-dimensional sections of these lattices should be drawn out as required for speedy reference. Three-dimensional models of the reciprocal lattices of the common crystal structures can be purchased (see List of suppliers), and a number of frequently encountered sections through these lattices are shown in Appendix 4.

It may be asked why it is that a particular section through the reciprocal lattice correlates so exactly with the diffraction pattern? The reason for this correlation is made clear by a consideration of the geometric construction, which carries the name of its originator, Ewald.

4.7.2 EWALD SPHERE CONSTRUCTION

Consider a set of planes of spacing, d and an electron beam striking them at an angle, θ, as drawn in a very exaggerated form in Fig. 4.20, since the actual values of θ in electron diffraction are 1°–2°. The Ewald sphere, of radius $1/\lambda$, is then drawn with its centre placed at a distance, $1/\lambda$, back along the incident ray, at I, where λ is the wavelength of the incident electron beam. Now a line from I drawn parallel to the diffracted beam OR, strikes the sphere at a point P, and encloses an angle $O\hat{I}P = 2\theta$. Ewald showed that the distance OP was equal to $1/d$, and that P was therefore a point on the reciprocal lattice, which has O as its origin, for the set of planes with interplanar spacing, d. So it is clear that this construction satisfies the Bragg equation,

$$2\theta = \frac{1/d}{1/\lambda}$$

or

$$\lambda = d\,2\theta = 2d\sin\theta$$

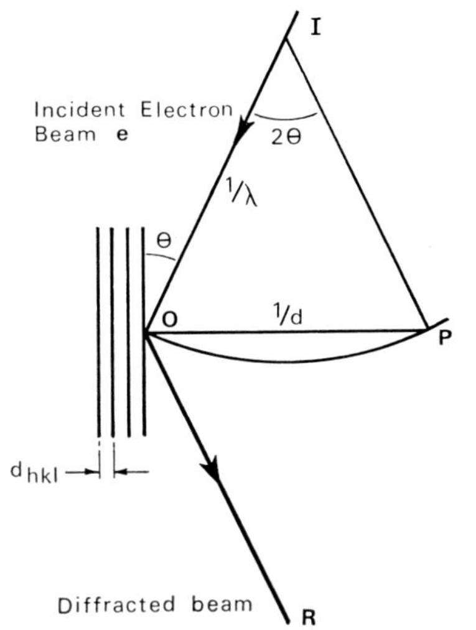

Fig. 4.20. Ewald sphere construction.

as

$$2\theta = 2 \sin \theta$$

for small angles θ, and that whenever the Ewald sphere intersects a reciproca lattice point, the Bragg equation will be satisfied, diffraction will occur, and a spot will appear in the diffraction pattern.

Of course, the crystal lattice is three-dimensional, the reciprocal lattice is in three dimensions, and the Ewald sphere is actually a sphere, centred at I, and not just the segment of a circle indicated in Fig. 4.20. However, the radius of the sphere is so large compared with the spacing between the reciprocal lattice points, that for a few points on either side of the origin, O, the sphere may be approximated to a plane. Also, the reciprocal lattice points are not infinitesimally small, and in fact, the points are extended to a length $2/\varepsilon_{hkl}$, in each [hkl] direction in the reciprocal lattice, where ε_{hkl} is the thickness of the specimen in that direction. So the reciprocal lattice points are streaked in the direction of thinness of the specimen; they are sometimes termed reciprocal lattice rods, or rel rods or spikes, being the longer the thinner the specimen. For example, in a thin f.c.c. crystal specimen with [002] zone axis, the (200) planes lie perpendicular to the specimen surfaces (Fig. 4.21a), and the 200 reciprocal lattice points are extended in the [002] direction, the direction of thinness of the specimen, as are all the other points in the [002] reciprocal lattice section in Fig. 4.21b.

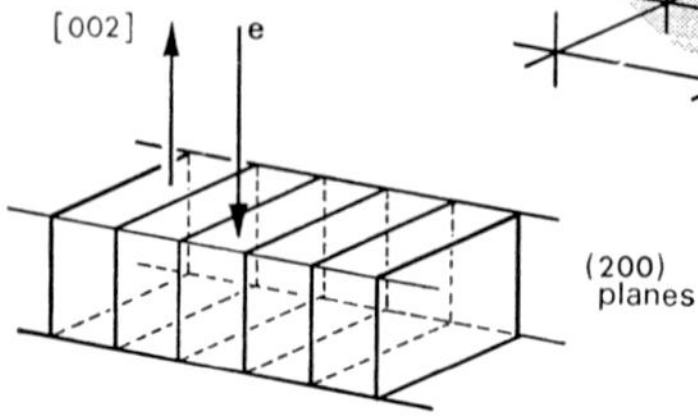

Fig. 4.21a. Thin foil of an f.c.c. specimen, oriented with the [002] zone axis anti-parallel to the incident electron beam.

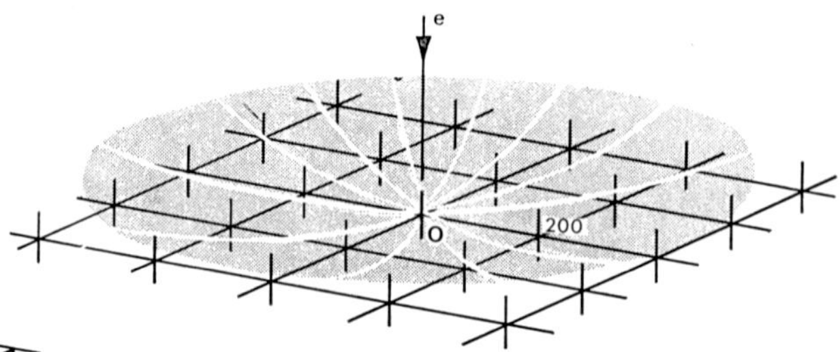

Fig. 4.21b. [002] Reciprocal lattice section with the Ewald sphere superimposed upon it.

If the Ewald sphere is now drawn symmetrically on the [002] reciprocal lattice section, it is clear that the sphere will intersect many reciprocal lattice rods, and therefore many spots will appear in the diffraction pattern, symmetrically spaced about the central spot, as in Fig. 4.9. However if the specimen is tilted slightly, then the Ewald sphere will no longer intersect reciprocal lattice rods symmetrically about the central points, and so the diffraction pattern will consist of an asymmetric array of spots. To some extent this asymmetry may be used to determine the angle of tilt of the specimen, from the symmetry position. However, specimens are often slightly buckled or curved, and so fine determination of specimen orientation by analysis of the asymmetry of the diffraction pattern may lead to spurious results. Fortunately, a further diffraction phenomenon, the formation of Kikuchi lines, which will be described in § 7.1, allows exact determination of the specimen orientation in any tilted position.

Of course, in the three-dimensional model of the reciprocal lattice, there are other sections, parallel to the [002] section in Fig. 4.18, both above and below, and if the Ewald sphere curves up sufficiently to intersect rel rods in the reciprocal lattice section above, then the far out spots in the diffraction pattern will come from lattice planes represented by points in this reciprocal lattice section. These diffraction spots will be indexed with the l index equal to 2, of the h, k, 2 type, whereas the close-in spots have indices of the type h, k, 0, with l always equal to 0 for this [002] zone axis.

Thus by using the reciprocal lattice and Ewald sphere constructions, many diffraction patterns are readily analysed, by examination of the appropriate reciprocal lattice model or of selected sections, and by consideration of the radius of the Ewald sphere and the extension of the reciprocal lattice points in the direction of thinness of the specimen.

4.8 Indexing the diffraction pattern from an unknown specimen

In § 4.2, procedures were described for indexing the diffraction pattern from a specimen of known crystal structure, for which the correct formula from Appendix 2a, relating the measured d-spacings and the indices $\{h, k, l\}$, was known. However, if the specimen and its crystal structure are completely unknown, the problem of indexing the diffraction pattern is very much more complex and no rigorous procedure can be laid down. A trial and error approach must be used, combined with some intuition and whatever limited knowledge of the specimen is available, to find the correct formula from Appendix 2a, which will give a consistent set of indices, which when put back into the associated formula for the angles ϕ from Appendix 2b, gives values which match the angles measured from the diffraction pattern. Working through the formulae in Appendix 2a, for all the possible combinations of the lattice parameters a, b and c, would be extremely time consuming. Fortunately this procedure may often be short-circuited if a number of diffraction patterns with different zone axes is available from the same specimen area. This generally involves the use of a specimen tilting stage, which is now available with most electron microscopes.

The symmetry of the diffraction patterns may give a indication of the structure of the specimen, for example

(a) an exactly hexagonal (60°) array of spots in the diffraction pattern will almost certainly indicate either an hexagonal or a cubic structure,

(b) a square (90°) array of spots will almost certainly come from a cubic, tetragonal or orthorhombic structure, and

(c) if no highly symmetrical pattern is present among a large number of patterns from the same specimen area, the structure is quite probably monoclinic or triclinic.

Also the existence of numerical relationships between the d-values, taken from the different diffraction patterns, may indicate a certain structure. The procedure is to determine all the different d-spacings from the spots in the various diffraction patterns, making due allowance for any changes in the camera constant, λL, which may occur during specimen tilting (see §§ 5.1.3 and 6.2.1), and to compile a list of $1/d^2$ values, in ascending order. Then by observation, establish whether any common factor or recognizable number sequence exists in this list of $1/d^2$ values (Andrews et al. 1971).

4.8.1 CUBIC STRUCTURES

The d-spacing formula for the cubic system, taken from Appendix 2a, gives

$$\frac{1}{d^2} = \frac{h^2 + k^2 + l^2}{a^2} = \frac{N}{a^2}$$

where

$$(h^2 + k^2 + l^2) = N.$$

Now the selection rules for cubic structures, given in § 4.5, indicate that for (i) the f.c.c. structure, h, k and l must all be even (where 0 is even). The permissible values from Table 4.1 are then

(hkl)	N
(111)	3
(200)	4
(220)	8
(311)	11
(222)	12
(400)	16
(331)	19
(420)	20
(422)	24
...	...

that is, $N = 4n$ or $8n + 3$, where n is an integer. Since $1/d^2 \propto N$, the number sequence to anticipate in the compiled list of $1/d^2$ values for the unknown specimen, in order to detect an f.c.c. structure is

$$3, 4, 8, 11, 12, 16, 19, 20, 24, \ldots,$$

or, of course, any multiple of these numbers. The practised eye easily recognizes this diffraction pattern, in particular, the sequence of rings in a polycrystalline pattern which will give $1/d^2$ values proportional to this series of N values, as for instance in the f.c.c. diffraction pattern in Fig. 1.7.

(ii) the b.c.c. structure, $(h + k + l)$ must be even, so $(h^2 + k^2 + l^2) = N$ must also be even. The permissible reflections then are

(hkl)	N
(110)	2
(200)	4
(211)	6
(220)	8
(310)	10
(222)	12
(321)	14
(400)	16
(411) or (333)	18
(420)	20
...	...

So the number sequence to look for among the compiled list of $1/d^2$ values for the unknown specimen, is the complete series of even numbers, for a b.c.c. structure to be present.

4.8.2 TETRAGONAL STRUCTURES

The d-spacing formula for the tetragonal structure is

$$\frac{1}{d^2} = \frac{h^2 + k^2}{a^2} + \frac{l^2}{c^2} = \frac{M}{a^2} + \frac{l^2}{c^2}$$

where $M = (h^2 + k^2)$ and can take values

$$1, 2, 4, 5, 8, 9, 10, 13, 16, 17, 18, 20, \ldots$$

A prominent ratio of 2:1 is found among these numbers, so if the compiled list of $1/d^2$ values shows this ratio predominantly, then the structure is likely to be tetragonal, and the above equation may be used to solve for $\{hkl\}$ and for a and c. The d-spacings from the $1/d^2$ values with a 2:1 ratio will be from the $\{hk0\}$ reflections, and these d-spacings are used to solve for h, k and a, while other d-spacings not fitting the 2:1 ratio pattern, will be from $\{hkl\}$ reflections and are finally used to solve for l and c.

4.8.3 HEXAGONAL STRUCTURES

The d-spacing formula for the hexagonal system is

$$\frac{1}{d^2} = \frac{4}{3}\frac{(h^2 + hk + k^2)}{a^2} + \frac{l^2}{c^2} = \frac{4}{3}\frac{P}{a^2} + \frac{l^2}{c^2}$$

where $P = (h^2 + hk + k^2)$ and can take the values

$$1, 3, 4, 7, 9, 12, 13, 16, 19, 21, \ldots$$

A prominent ratio of 3:1 is found among these values, so if the compiled list of $1/d^2$ values shows this ratio predominantly, then the structure will probably be hexagonal and the above equation may be used to solve for $\{hkl\}$ and a and c. As before, the d-spacings with the prominent 3:1 ratio will be from the $\{hk0\}$ reflections and can be used to solve for h, k and a, while the other d-spacings will be from $\{hkl\}$ reflections and are used finally to solve for l and c.

4.8.4 GENERAL CASE

As the symmetry of the structure becomes less pronounced, the analysis of the measured d-spacings to determine the specimen structure and thus index the diffraction spots or rings in the recorded patterns becomes increasingly difficult and time consuming. Ito (1949) has given a general treatment, which may provide correct solutions, but is limited by the fact that planes with low Miller indices may not give diffraction spots for structures of low symmetry under certain conditions.

Another approach, when the previous procedures fail to provide a solution, is to search through the A.S.T.M. index to find a material with exactly similar d-spacings to those determined for the unknown specimen. As electron diffraction in the electron microscope does not generally give very accurate values for the d-spacings, a number of possible materials with approximately similar d-spacings may be indicated. Here any corroboratory evidence, to eliminate some of the possibilities, should be used.

If sufficient specimen material is available, X-ray diffraction patterns will give much more accurate d-spacings, and probably enable an accurate analysis of the material. Also X-ray micro-analysis will reveal the elemental constituents of the specimen, and reduce the choice of possible alternative materials very considerably. If the crystalline specimen area is only on a microscopic scale, an instrument allowing simultaneous X-ray micro-analysis, high magnification observation and electron diffraction, such as the AEI EMMA-4, is almost certainly required.

If a precise analysis of the unknown material is possible by this procedure, the A.S.T.M. index will reveal the structure of the specimen and its lattice parameters. The appropriate formula from Appendix 2a will then permit the assignment of the correct indices to the diffraction spots or rings in the various diffraction patterns recorded.

REFERENCES

Andrews, K. W., D. J. Dyson and S. R. Keown (1971), Interpretation of electron diffraction patterns (Hilger and Watts Ltd., London).
A.S.T.M. Index, Index to the powder diffraction file (Am. Soc. for Testing Mater., Philadelphia, U.S.A.).
Ito, T. (1949), A general X-ray powder photography, Nature *164*, 755.
Roser, W. R. and G. Thomas, (1964), Slide rule method for indexing electron diffraction patterns, Rev. Sci. Instr. *35*, 613.

FURTHER READING

Quinn, T. B. (1970), X-rays, electrons and crystalline materials (Butterworths and Co. Ltd., London).
Rymer, T. B. (1970), Electron diffraction (Methuen and Co. Ltd., London).

Chapter 5

The formation of a useful diffraction pattern from a specimen in the electron microscope

The formation of the electron diffraction pattern in the back focal plane of the objective lens was described in Chapter 3. The particular *mode of operation* of the microscope used to transfer the diffraction pattern on to the viewing screen, and thence on to the photographic plate is governed by the *type of specimen* under examination, and the *particular microscope* being used.

Five diffraction *modes of operation* of the electron microscope are generally recognised,

(i) selected area diffraction
(ii) low angle diffraction
(iii) high resolution diffraction
(iv) reflection diffraction
(v) scanning electron diffraction.

The particular mode of operation will be determined by the *type of specimen* from which diffraction information is required, and will be discussed separately under the five modes of operation. In outline the corresponding specimens are

(1) metals, minerals and materials with lattice spacings smaller than 2.5–4.0 nm

(2) materials, polymers and organic crystals with lattice spacing greater than 2.5–4.0 nm

(3) metals and materials, generally in an impure state, giving diffraction spots or rings so close together in the diffraction pattern that they cannot be resolved in (1)

(4) solid specimens or specimens coated with a surface layer, perhaps of an oxide

(5) specimens undergoing dynamic transitions, due to heating, cooling, straining, or thin film growth; or when quantitative measurements of the intensities of the diffraction spots are required.

The actual operating procedure for the different makes of electron microscope are described extensively in the manufacturers' operating manuals, and no attempt will be made to reproduce them here, except in general terms applicable to the majority of instruments.

First then, the standard operating diffraction mode used for the study of the majority of metals, minerals and materials will be described.

5.1 Selected area diffraction

The great advantage of selected area diffraction in the electron microscope is the very small volume of the specimen from which a diffraction pattern can be formed, and structural information thus obtained. An electron diffraction pattern is easily formed from an area 1 μm × 1 μm, of a specimen of thickness 50 nm – a total volume of 5×10^{-14} cm^3 – after a photographic exposure of only a few seconds. By comparison, the minimum area selectable by X-ray or other diffraction technique is between 20 and 100 × larger than this, and still requires very long exposure times. How then is this very small area of a thin specimen selected?

5.1.1 INSTRUMENTAL DESIGN CONSIDERATIONS

Consider again the formation of the diffraction pattern in the back focal plane of the objective lens (Fig. 5.1). Now by placing a very small aperture either just above or just below the specimen plane AB, the area from which the pattern is taken could be selected, by first viewing the image and the aperture together in the normal magnification mode, centering the aperture around the area of interest, and then switching the microscope to the diffraction mode. However selection of an area 1 μm in diameter is desired, and such small apertures are very difficult to produce and to keep clean, and in addition, there is insufficient space between the pole-pieces of the objective lens close to the specimen to insert such an aperture in a high resolution electron microscope. A high resolution objective lens requires a very small gap between the lens pole-pieces, just sufficient to insert the specimen, the objective aperture and perhaps two anti-contaminator blades.

Fortunately there is another plane in which it is possible to insert the area-

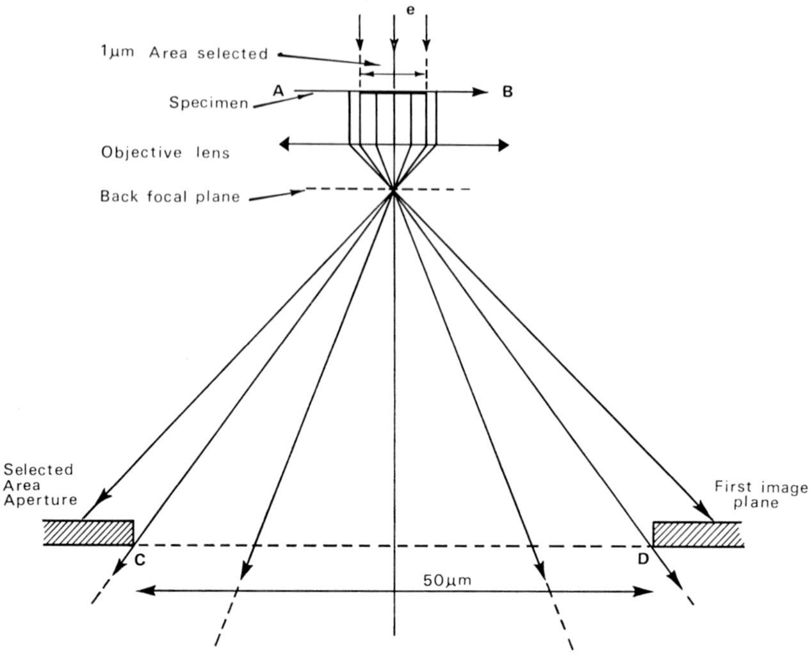

Fig. 5.1. Ray diagram showing the area of the specimen selected by an aperture positioned in the first image plane of the objective lens, assuming simple geometric optics.

selecting aperture, overcoming these difficulties and introducing only one other problem, that of inaccuracy in the selection of the area. This preferred plane is the image plane CD (Fig. 5.1) of the objective lens, at which position the specimen area is magnified approximately 50 × (this value may vary between 30 and 100 × depending upon the type of microscope being used), and so the area-selecting aperture can be 50 × larger than the area which it is required to select. 50 μm apertures are easily produced, easily kept clean, and their insertion in the image plane of the objective lens presents few design difficulties.

All modern high resolution electron microscopes have a mechanism for inserting area-selecting apertures of different sizes in the image plane, CD, of the objective lens, and controls for centering the aperture around the area of interest while viewing the image. The one problem in selecting a particular area of the specimen in this manner is in ensuring that the area of the image selected actually corresponds to the area of the specimen from which the diffraction pattern is obtained. Considerable inaccuracies in area selection may occur (even though the aperture is centered around the area of interest in the image), unless a strict procedure is followed. This is partially automatic in some microscopes, and requires manual adjustment in others.

5.1.2 THE MODE OF OPERATION OF THE MICROSCOPE

(a) The first projector lens, P_1, (the lens following the objective lens, sometimes called the intermediate or the diffraction lens) is set to focus on the plane of the area-selecting aperture, by first inserting the aperture and then adjusting P_1 until the image of the aperture is in focus on the viewing screen. In some microscopes this is achieved automatically when a particular magnification is selected–indicated, selected area (SA) magnification–or when a particular selected area magnification range is used. In other microscopes manual setting of the P_1 control is required. The assistance of the relevant instruction manual should be sought here.

(b) With P_1 focused on the plane of the area-selecting aperture the image of the specimen is now focused into the same plane by adjusting the objective lens FOCUS controls until the final image comes into focus on the viewing screen. This is always a manual operation.

(c) The microscope is switched to the diffraction mode, by depressing the relevant button, or by turning the relevant switch, or by reducing the strength of P_1 until the diffraction pattern comes into focus on the viewing screen, as instructed in the manual. Where a button or switch must be operated, a separate diffraction pattern focus control, which only operates in the diffraction mode, is generally provided for fine adjustment of the focus of the diffraction pattern.

The diffraction pattern can now be studied on the viewing screen or photographed in the usual way–remember to defocus the second condenser lens, C_2, in order to achieve a sharp diffraction pattern and the correct photographic exposure. The objective aperture which will normally be in place during operation in the imaging mode, surrounding the central diffraction spot, must be removed in order to observe the full diffraction pattern. The procedure for recording the diffraction pattern is described in § 3.3.

5.1.3 CALIBRATION OF THE CAMERA CONSTANT, λL

In order to analyse the diffraction pattern, as described in § 4.2, it is necessary to have an accurate value of the camera constant, λL. As indicated previously, it is often insufficient to take predetermined values of λ and L, as both these values may alter, either purposely due to adjustments to the high voltage supply or the lens currents, or unintentionally because of instabilities in the electronic circuitry. So from one experiment to the next, the value of λL may vary as

$$\frac{\delta\lambda L}{\lambda L} = \frac{\delta\lambda}{\lambda} + \frac{\delta f_0}{f_0} + \frac{\delta M_{P_1}}{M_{P_1}} + \frac{\delta M_{P_2}}{M_{P_2}},$$

where f_0 is the focal length of the objective lens and M_{P_1} and M_{P_2} are the magnifications of the P_1 and P_2 lenses respectively. $\delta\lambda/\lambda$ is less than 10^{-5} in a modern high resolution instrument, but the other terms in the equation may be considerably larger, depending upon hysterisis effects when the lens currents are adjusted. Some instruments have a lens current standardization control to prevent inaccuracies in the magnification of the projector lenses due to hysterisis. The remaining inaccuracies will then be due to lens current supply instabilities, which should be less than 5×10^{-6} for the objective lens and 10^{-4} for the projector lenses.

However, for an accurate analysis of the diffraction pattern, particularly of an unknown material or of a specimen undergoing a non-axis-centred tilting experiment, when f_0 must be altered in order to focus on different areas of the tilted specimen, it is preferable to calibrate the camera constant for each diffraction pattern recorded. This is generally achieved by coating the specimen with a thin layer of a polycrystalline material, if the specimen is a single crystal, or with a single crystal, if the specimen is polycrystalline (§§ 6.2.1 and 6.2.2).

5.1.3a *Thallous chloride coating*

For general diffraction purposes, vacuum-deposited thallous chloride forms an excellent calibrating film, giving a polycrystalline diffraction pattern of many sharp, diffraction rings, similar to Fig. 1.7. Thallous chloride has a simple cubic structure, for which the lattice parameter, $a = 0.3842$ nm (Hambling 1953) is independent of evaporation conditions or exposure to the atmosphere. The thin polycrystalline film may either be coated directly on to the specimen before examination, or after an area of interest has been found by observation in the microscope. The specimen may be coated while mounted in the specimen holder, a simple mask being used to protect the specimen holder yet allow deposition on the specimen during vacuum evaporation in the coating unit. A satisfactory coating of thallous chloride is readily achieved using the following conditions:

Material: 20 mg TlCl crystals (see List of suppliers).

Method of Evaporation: TlCl crystals placed in a molybdenum boat.

Source-to-specimen distance: Approx. 120 mm.

Vacuum: Better then 10^{-5} torr.
Evaporation time: Until TlCl crystals completely evaporated.
Structure: Cubic.
Lattice parameter, a: 0.3842 nm

Two precautions must be taken when using thallous chloride for this purpose,

(i) thallous chloride is a TOXIC material, and care must be taken to avoid inhalation, and

(ii) the thallous chloride coating may be evaporated from the specimen by high electron beam intensities, so care should be taken to avoid this condition occuring.

It is then a simple matter to index the cubic diffraction pattern and determine the value of λL, using the procedure described in § 4.2, the d_{hkl} values being determined using the cubic formula from Appendix 2a, and $a = 0.3842$ nm, and the R values being measured from the diffraction pattern.

5.1.3b *Aluminium and gold coatings and calibration specimens*

For less accurate determination of the camera constant, perhaps when the specimen identity is known and diffraction analysis is required only for orientation determination, as described in § 6.1, a coating of vacuum-deposited aluminium may be used. An aluminium coating is rather more stable than thallous chloride under the electron beam, but its lattice parameter is somewhat dependent on evaporation conditions. Alternatively a thin film of aluminium or gold may be sandwiched next to the specimen using two specimen grids. The film is deposited on to a clean substrate, usually freshly-cleaved mica, from which it is then floated on to water, collected and cut into small pieces for positioning on the grids next to the specimen. Satisfactory films of aluminium and gold may be prepared using the following conditions,

Aluminium Evaporation

Material: 25 mm of 0.25 mm diam. wire or 10 mm of 15 mil. diam. wire (see List of suppliers).
Method of Evaporation: Al wire wrapped around V-shaped tungsten filament.
Source-to-substrate/specimen distance: 120 mm.
Vacuum: 5×10^{-5} torr.
Evaporating time: Until Al wire completely evaporated.
Structure: Face-centred Cubic.
Lattice parameter a: 0.4041 nm.

Gold Evaporation

Material: 10 mm of 0.25 mm diam. wire or 15 mm of 8 mil. diam. wire (see List of suppliers).

Twice these lengths should be used for evaporation on to a substrate rather than a specimen or carbon support film.

Method of evaporation: Au wire wrapped around V-shaped tungsten filament.

Source-to-substrate/specimen distance: 120 mm.

Vacuum: 5×10^{-5} torr.

Evaporating time: Until Au wire completely evaporated.

Structure: Face-centred cubic.

Lattice parameter a: 0.4070 nm.

Diffraction calibration specimens are available from a number of accessory suppliers (see List of suppliers).

5.1.3c. *Calibration of the camera constant against objective lens current*

Alternatively a calibration chart may be drawn up showing the variation in the camera constant with the current in the objective lens. The calibration procedure consists of forming successive diffraction patterns from a standard calibration specimen at various specimen heights, by tilting an off-axis area of the specimen through different tilt angles using a specimen tilt holder. The camera constant is then calculated for each diffraction pattern and plotted on the calibration chart for the corresponding objective lens current. If a reading of the objective lens current is then recorded following a diffraction pattern exposure, a correction factor may be applied to the camera constant to allow for any variation in specimen height (Rickards and Williamson 1964; Armitage and MacConaill 1964). This procedure requires a more accurate objective lens current meter than that generally supplied with a standard instrument and also frequent recalibration to detect possible changes in the accelerating voltage or projector lens currents, if the procedure is to be used for accurate analyses.

5.1.3d *Internal calibration standards*

Some specimens have 'built-in' calibration features. For example, the matrix of a known metal foil may serve as the calibrating standard for diffraction patterns obtained from unknown precipitates in the specimen.

5.1.4 THE MINIMUM SPECIMEN AREA WHICH CAN BE SELECTED ACCURATELY

Apertures as small as 5–10 μm in diameter are available and might be used as area-selecting apertures in the first image plane. With a magnification of 50 × between the specimen and the first image plane, it might be expected that areas of the specimen as small as 0.1 μm–0.2 μm in diameter could be selected using these apertures. However, even if the procedure for bringing the first image and the area selecting aperture into exactly the same plane is followed rigorously, a further inaccuracy in area selection occurs, due to the spherical aberration of the objective lens. This cannot be eliminated by the operator, and makes the inaccuracies in area selection such that it is not feasible to select an area smaller than 1 μm in diameter, with reasonable accuracy. Agar (1960), Phillips (1960) and Reicke (1961) have shown that the inaccuracy between the area of the image selected by the aperture, and the area from which the diffraction pattern is actually formed, is given by the term $C_s\alpha^3$, if no focusing errors are present, where C_s is the spherical aberration coefficient of the objective lens, and α equals 2θ, where θ is the Bragg angle for each of the various diffraction phenomena in the diffraction pattern. Typical values of C_s and α are 3 mm, for modern high resolution microscopes, and 0.02 rad, for low order reflections in a typical metal specimen at 100 kV, respectively.

Neglecting focusing errors, the inaccuracy of selection for these typical values is

$$C_s\alpha^3 = 3 \times 0.02^3 = 0.24 \text{ μm}$$

that is, a 24% error in selection of an area of 1 μm diameter. Of course, for higher order reflections, the values of α will be larger and the inaccuracy of area selection $C_s\alpha^3$ correspondingly much greater.

A particular advantage of high voltage electron microscopy is that the values of α are reduced by a factor of 4 × at a 1000 kV over those at 100 kV, and α^3 by a factor of 64 ×, so that it is potentially possible to select an area approximately 15 nm in diameter at 1000 kV, with the same accuracy as a 1 μm diameter area can be selected at 100 kV.

Inaccuracies in image focus almost always occur, often by as much as 5 μm out-of-focus for a thick specimen, and give rise to a further inaccuracy in area selection, $D\alpha$, where D is the amount out-of-focus. Typically,

$$D\alpha = 5 \times 0.02 = 0.1 \text{ μm}$$

for low order reflections, making the total inaccuracy of area selection

$$y = D\alpha + C_s\alpha^3$$

such that an area less than 1 μm in diameter, corresponding to a 50 μm area selecting aperture, cannot be selected with any reasonable accuracy at 100 kV. The equivalent value at 1000 kV is approximately 0.1 μm, assuming a similar inaccuracy in focusing.

5.1.5 SPECIMENS FROM WHICH USEFUL SELECTED AREA DIFFRACTION (SAD) PATTERNS CAN BE OBTAINED

Crystalline areas of metals, minerals and materials of all types, with lattice or periodic spacings, d, smaller than approximately 2.5 nm, can be expected to give a useful SAD pattern. The materials may be organic crystals, polymers or crystalline inclusions in biological tissue or in other amorphous specimens. Specimens with spacings greater than 2.5 nm will not give a useful SAD pattern as can be seen from a consideration of the basic formula for the analysis of diffraction patterns (see § 4.1)

$$Rd = \lambda L.$$

R is the distance between the central diffraction spot and other diffraction spots or rings, and should be no less than ~4 mm so that an accurate measurement can be made and the spots are not lost in the diffuse scattering around the central spot. d is the lattice spacing, λ is the electron wavelength, 0.004 nm at 80 kV, and L is the diffraction camera length, which may vary between a minimum of ~250 mm and a maximum of ~2500 mm, depending upon the instrument being used. It is the maximum length of 2500 mm which is important here. Substituting the values for R and L, the maximum value of d, which will give a satisfactory SAD pattern, is

$$d = \frac{0.004 \times 2500}{4} = 2.5 \text{ nm}.$$

This figure may be extended to perhaps 4.0 nm if the minimum acceptable value of R is made 2.5 mm, by taking care to reduce the scattering around the central spot using a diffraction beam stop and careful photographic exposure.

However, many specimens, particularly organic crystals, polymers and biological crystals, have lattice spacings, d, much in excess of 4.0 nm. For instance, a beef liver catalase crystal shows a cross lattice with 6.85 nm and 8.75 nm lattice spacings (Wrigley 1968), and other crystalline specimens have spacings approaching 100 nm; ferritin molecules crystalize in an array of

~10 nm spacing, collagen fibrils have a spacing 69 ± 1 nm (Ferrier and Murray 1966), while the spacing of a diffraction grating replica (21,600 lines/cm) is 463 nm.

To form a useful diffraction pattern from these specimens it is necessary to use low angle diffraction techniques, as discussed in the following section.

5.2 Low angle diffraction

5.2.1 INSTRUMENTAL DESIGN CONSIDERATIONS

It is clear from the discussion in § 5.1.5 that in order to achieve a useful diffraction pattern from specimens with large values of d corresponding to small or 'low' Bragg angles, while maintaining R greater than 2.5 mm, for ease of observation and measurement, it is necessary to markedly increase the value of either λ or L. λ is the wavelength of the incident electron beam and can only be increased slightly by decreasing the accelerating voltage, which however should be avoided, as this will decrease specimen penetration and increase chromatic blurring of the image. It is therefore necessary to increase the value of L, the diffraction camera length. The value of L, in the standard mode of operation is given by

$$L = f_0 \times M_{\mathrm{P}_1} \times M_{\mathrm{P}_2} (\times M_{\mathrm{P}_3}) \quad \left(\begin{array}{l}\text{if the instrument has}\\ \text{three projector lenses.}\end{array}\right)$$

where f_0 is the focal length of the objective lens, and M_{P_1}, M_{P_2} and M_{P_3} are the magnifications of the projector lenses. The focal length of the objective lens, f_0, is generally 2–5 mm, and with the maximum magnifications of the projector lenses that can be used in the diffraction mode, the value of L is limited to approximately 2500 mm. It is clear then that the main limitation to increasing the value of L, is the very small value of f_0. If the value of f_0 is increased by reducing the excitation of the objective lens with a coarse focus control, the area selecting facility is lost, as the image of the specimen is no longer focused in the plane of the area-selecting aperture. If the area selecting facility must be retained the distance of the specimen from the objective lens must be increased, as the lens strength is weakened. However the lens is now working in a far from optimum state and the attainable image resolution is poor. In order to overcome this difficulty, special specimen stages have been constructed (Ferrier and Murray 1966; Yeh and Geil 1967) for microscopes with large specimen chambers in which the specimen can be raised so as to increase f_0 to give a long camera length for diffraction, and then lowered to the normal specimen plane for recording an image of high quality, the diffracting area being observed continuously during the operation. In this way, the camera length, L, can be increased by a factor of approximately 5 ×. This will give an increase of 5 × in the maximum lattice spacing, d, which can be studied, with an area selecting facility. However a greater increase in camera length is still required for many specimens, containing very large d-spacings, and in addition, this type of specimen stage can only be fitted to a limited number of microscopes and is not generally commercially available.

If specimens with very large d-spacings are to be studied by diffraction, other techniques for increasing the diffraction camera length must be used, foregoing the facility to select a specific 1 µm area of the specimen for diffraction analysis. This will not be detrimental if there is only one type of crystal in the general area of the specimen, but may introduce difficulties if two or more types of crystal are in close proximity and it is necessary to discriminate between them.

5.2.2 MODES OF OPERATION OF THE MICROSCOPE

Four different modes of operation of the electron microscope for low angle diffraction studies are generally available. As some instruction is usually given in the relevant instruction manual, no attempt will be made to give the operating procedure for the various instruments in detail, but the principles of the different techniques will be described.

A difference in the actual operating procedures occurs between instruments with three-lens and four-lens imaging systems, but as this does not affect the principles of operation, only the three-lens imaging system (objective and two projector lenses) will be discussed.

The choice of one of the four available modes of low angle diffraction operation depends primarily on the specimen under examination, in particular, the magnitude of its lattice spacings. In addition, consideration must be given to the amount of detail expected in the diffraction pattern and the proximity of the features in the specimen to one another, since these factors affect the resolution required in the diffraction pattern. Little advantage is gained in increasing the camera length if the diffraction phenomena expand equally with the distances between them. In order to observe and measure diffraction spots that are very close to each other and to the central spot, it is necessary to increase the camera length without expanding the size of the spots or the width of the rings. Therefore the resolution in the diffraction pattern must be increased, in addition to increasing the camera length.

In order to compare the various modes of low angle diffraction operation, a criterion for resolution must first be established. Here it will be assumed that two diffraction spots can just be resolved if the separation of the spot centres is equal to the diameter of either of the spots. On this criterion the minimum angular resolution, α_{min}, will be given by the diameter, D, of the central spot on the photographic plate formed by the undeviated electrons, divided by the camera length, L.

$$\alpha_{min} = D/L.$$

Now it is possible to compare the various modes of low angle diffraction operation in terms of the range of camera lengths and the angular resolutions available.

5.2.2a *Three-lens method*

In this mode of operation (Bassett and Keller 1964; Ferrier and Murray 1966; Ferrier 1967) the specimen is mounted in its normal position in the microscope and the objective and first projector lenses are switched OFF. The ray diagram for this mode of operation is shown in Fig. 5.2. The first condenser lens, C_1, is set to maximum excitation to give the smallest possible spot size, and the second condenser lens, C_2, set to image this spot in the object plane of the second projector lens, P_2, the diffraction pattern of the specimen being formed in this plane. To achieve the correct setting of C_2, the lens current is reduced until the sharpest image of the central spot is observed on the viewing screen, after magnification by the second projector lens. The camera length is now given by

$$L = y \times M_{P_2},$$

where y is the distance from the specimen to the plane of the initial diffraction pattern. The camera length can be varied by altering the strength of the P_2 lens; values in the range 10–100 m can generally be obtained. The corresponding d-spacings are 10–100 nm, on the previous criterion that $R = 4$ mm. For $R = 20$ mm, a distance which will still include the most important phenomena of the diffraction pattern on the photographic plate, d-spacings of 2–20 nm can be analysed. Thus a diffraction technique overlapping with the selected area mode is available, and between them they can cover most of the crystalline materials encountered. However, a number of disadvantages in this mode of low angle diffraction should be noted.

(i) In order to form a high resolution image of the specimen the objective and P_1 lenses must subsequently be excited, and some realignment of the

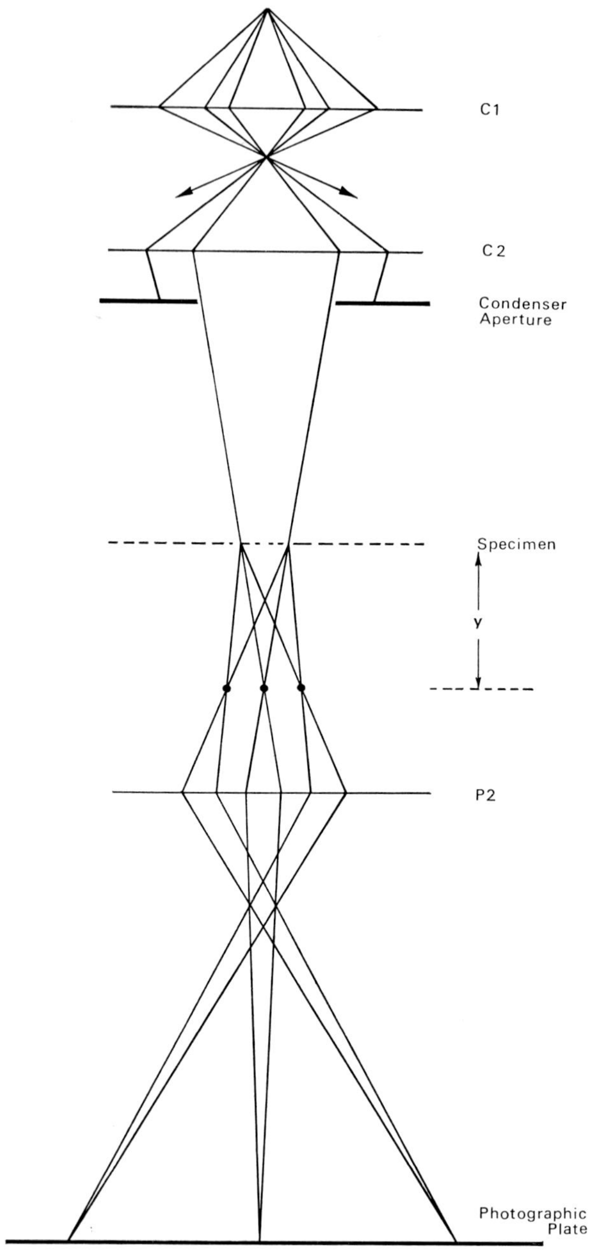

Fig. 5.2. Ray diagram for the three-lens mode of operation for low angle electron diffraction.

microscope column may be necessary, and consequently considerable difficulty in maintaining the area of interest may be experienced. For the recommended procedure for alleviating this difficulty, see Glaeser and Thomas (1969).

(ii) In consequence of (i), accurate area selection for diffraction analysis is not possible. The diameter of the area from which the diffraction pattern is taken is defined by the size of the condenser aperture, as seen in Fig. 5.2. In order to minimize this area, and also to reduce the diffuse scattering around the central spot of the diffraction pattern, very small condenser apertures may be used. Glaeser and Thomas (1969) have used 10–15 μm apertures. By increasing the strength of the C_2 lens, an image at low magnification and resolution may be observed on the viewing screen, and the objective aperture may be used to make a rough area selection; a subsequent decrease in the C_2 current brings the diffraction pattern back into focus again. With this arrangement, some of the diffracted beams from regions close to the edge of the 'selected' specimen area will not appear in the diffraction pattern.

(iii) In certain circumstances, electrostatic charging of the specimen may cause distortion in the diffraction pattern. With non-conducting biological specimens it is essential to coat the specimens lightly with carbon, or use a conducting support film, to prevent distortion of the diffraction pattern and possible damage to the specimen.

Distortion will cause a more serious error in measurements taken from the low angle diffraction pattern than from a selected-area pattern as the spots or rings are now effectively much closer together.

(iv) The angular resolution in the three-lens mode of operation is typically only $\alpha_{min} = 3 \times 10^{-5}$ rad (Ferrier 1967), compared with $\alpha_{min} = 4 \times 10^{-6}$ rad for selected area diffraction, and this may be insufficient to resolve fine detail in the diffraction pattern of some specimens.

For higher angular resolution then, other low angle diffraction modes must be used.

5.2.2b *Four-lens method*

In the four-lens mode of operation only the first projector lens is not excited, and again the specimen is loaded in the normal specimen position. The illuminating conditions are identical to those for the selected area mode, as shown in Fig. 5.3. The objective lens is weakly excited to focus the diffraction pattern in the object plane of P_2, the strength of P_2 being altered to vary the camera length. As the illuminating conditions are identical to those in the selected

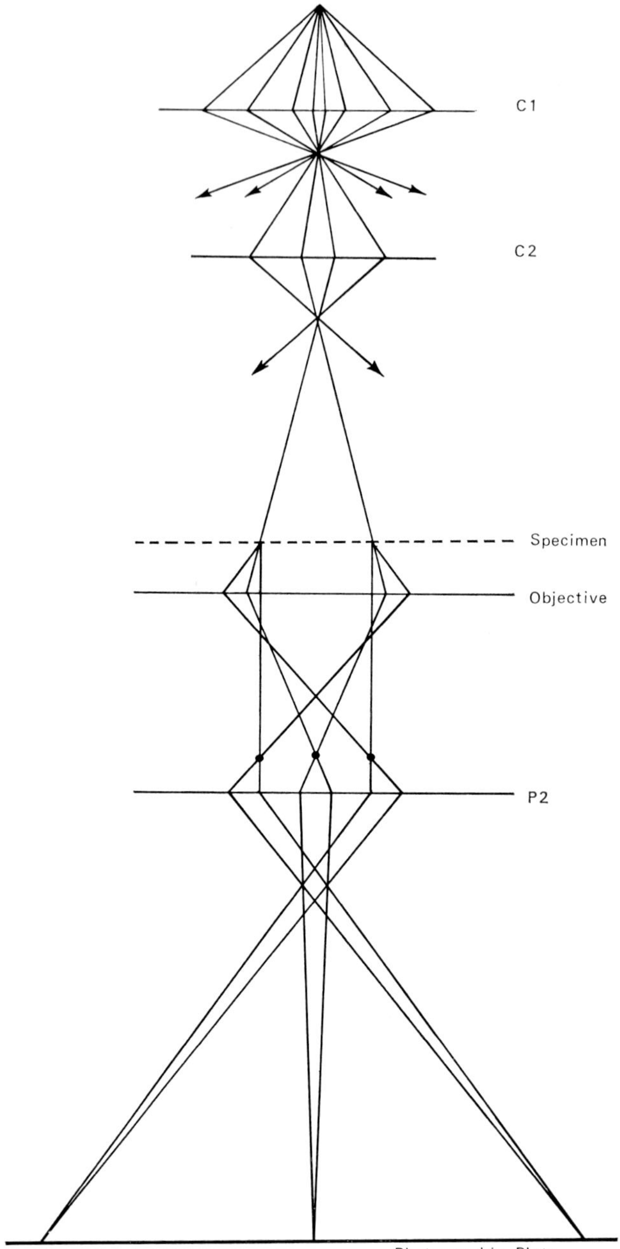

Fig. 5.3. Ray diagram for the four-lens mode of operation for low angle electron diffraction.

area mode, so the angular resolution is typically, $\alpha_{\min}$ 4×10^{-6} rad (Ferrier 1967). This mode has the further advantage that an image at magnifications up to about 20,000 × may be formed without realignment, by slowly increasing the strength of the objective lens.

Camera lengths up to 75 m can generally be obtained in this mode.

5.2.2c *Five-lens method*

A wider range of camera lengths can be obtained by using both the projector lenses to project the diffraction pattern on to the viewing screen. In this five-lens mode of operation the specimen is loaded in the normal position and the objective lens is weakened to focus the diffraction pattern in the object plane of P_1, the plane in which the first image is formed in the normal imaging conditions. Both of the projector lenses may now be varied to give a very wide range of camera lengths. The value of the camera length is simply given by

$$L = f_0 \times M_t,$$

where f_0 is the standard focal length of the objective lens and M_t is the total magnification of the microscope at any one particular setting. Thus if, for example, the focal length of the objective lens, f_0 is 3 mm, and the magnification range of the microscope is 1000–250,000 ×, then a range of camera lengths 3–750 m is available.

If the range of objective lens control on the microscope does not allow the objective lens to be weakened sufficiently to focus the diffraction pattern in the object plane of P_1, the objective lens may be switched off altogether. A similar range of camera lengths is available with this lens arrangement, though now, the second condenser lens must be focused into the object plane of P_1, in order to form the diffraction pattern in that plane, and this may seriously limit the resolution available in the diffraction pattern.

For diffraction patterns containing very fine detail, a greater angular resolution may be required, and here a fourth low angle diffraction mode of operation may be used.

5.2.2d *Ultra-high resolution method*

In this mode of operation the specimen is mounted in the plane of the area selecting aperture, either on the area selecting aperture mechanism or in a special specimen stage, and the two condenser lenses and the objective lens are set at maximum excitation to form an extremely small illuminating spot above the specimen (Curtis et al. 1967; Ferrier 1967). The ray diagram in

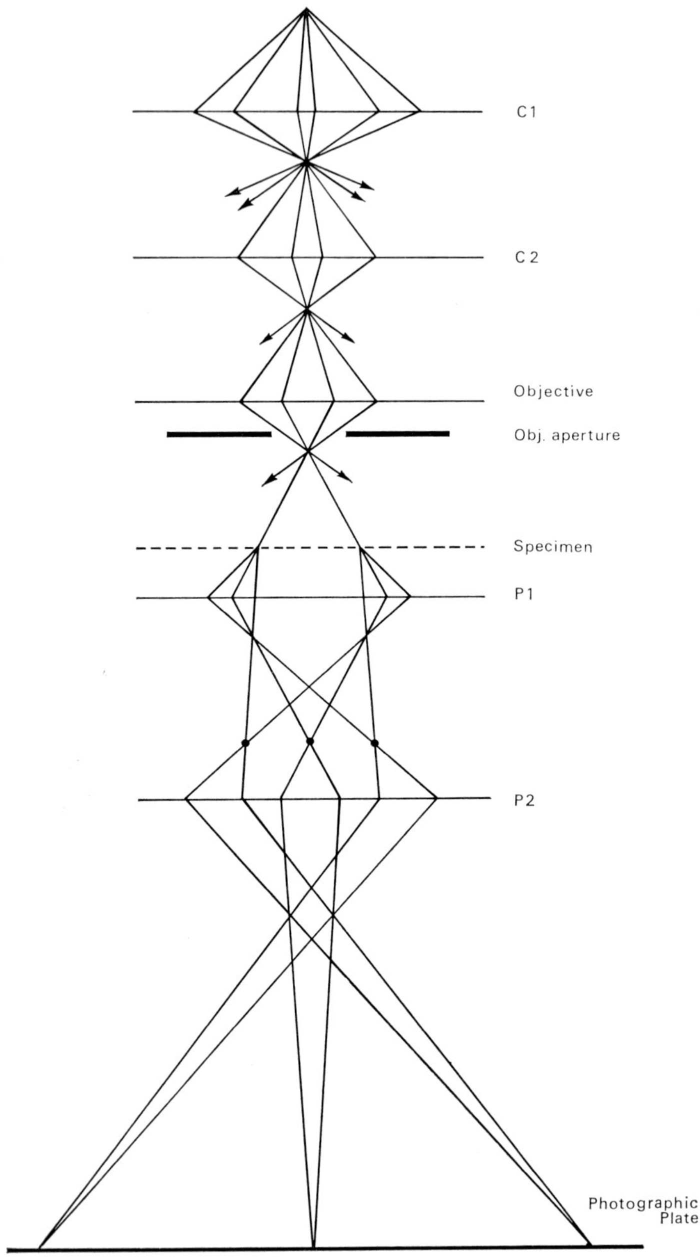

Fig. 5.4. Ray diagram for the ultra-high resolution mode of operation for low angle electron diffraction.

Fig. 5.4 shows that the P_1 lens must be adjusted so that the diffraction pattern is first formed in the object plane of the P_2 lens, which then focuses it on to the viewing screen. It is no longer possible to form a high resolution image with the specimen in this position, but a low magnification image at approximately 2000 × may be obtained by increasing the strength of the P_1 lens.

Typically angular resolutions of $\alpha_{min} = 1 \times 10^{-7}$ rad are possible in this mode, but a number of difficulties should be noted.

(i) The illumination spot above the specimen is now very small and therefore subject to astigmatism and spherical aberration. Astigmatism may be eliminated by using the standard condenser and objective astigmatism correctors of the microcope, and spherical aberration may be reduced to negligible proportions by limiting the specimen area contributing to the diffraction pattern, by placing a small aperture in contact with the specimen; Curtis et al. (1967) used a 400 μm aperture in this mode.

(ii) Electrons passing through the objective lens at wide angles may pass around the outside of the diffraction aperture mechanism, and be focused by the P_1 lens into the diffraction pattern, thus causing considerable confusion. If this does occur a further aperture should be included somewhat below the normal objective aperture, perhaps mounted from the specimen stage for ease of centering (Curtis et al. 1967), to prevent these rays reaching the final diffraction pattern.

(iii) Only two lenses are used to focus and magnify the diffraction pattern, and the available camera lengths are rather shorter than is desirable to take advantage of the increased angular resolution. It is therefore necessary to use the finest photographic emulsions, and due to difficulty in focusing the diffraction patterns, take through-focal series, in order to achieve ultra-high resolution in the diffraction pattern.

5.2.3 CALIBRATION OF THE CAMERA CONSTANT, λL

Although the approximate values of the available camera lengths in the selected area mode of operation are generally indicated on the control panels of commercial electron microscopes or in calibration charts, this is usually not the case for low angle diffraction modes. It is necessary then to determine the camera constant for each particular set of lens conditions, in order to analyse the diffraction pattern.

A range of standard specimens of known *d*-spacing is required coinciding as closely as possible with the unknown *d*-spacings which might be encountered.

Standard specimens for accurate calibration of the camera constants in the selected area mode are readily available, although some precautions must be taken in their preparation and storage, as indicated in § 5.1.3. However, specimens with large *d*-spacings rarely have the very highly ordered crystalline arrays of metals and mineral crystals, and therefore do not give very precise calibration of low angle diffraction camera constants; also they are very dependent upon the preparation technique and manner of storage.

Calibration specimens for low angle diffraction are therefore limited to a small number of readily available crystalline materials, for which a standard preparation technique has been established.

5.2.3a *Negatively-stained beef liver catalase crystals* – 8.75 nm × 6.85 nm ± 0.1 nm (Wrigley 1968)

The standard preparation technique for catalase calibration specimens was developed by Wrigley (1968). A solution of beef liver catalase (available from Boehringer and Soehne, see List of suppliers) in 11 % NaCl is dialized against decreasing concentrations of phosphate buffer at pH 6.3, ending either with 0.001 M buffer or with distilled water. A drop of the crystal suspension is placed on a carbon or formvar coated grid and allowed to almost dry. Just before the specimen becomes dry a drop of 4 % sodium silicotungstate negative stain is added. The stain is left on the grid for one minute, after which it is drained off as thoroughly as possible so that 'negative stain' does not build-up along the edges of the crystals.

Specimens are then coated with a light but clearly visible layer of evaporated carbon to prevent specimen charging in the microscope. This step is particularly important if good low angle diffraction patterns of crystalline catalase are to be obtained.

The high magnification image of a catalase crystal shows a rectangular lattice, with orthogonal spacings, $d_1 = 8.75$ nm and $d_2 = 6.85$ nm, and a diffraction pattern of the crystal has an orthogonal array of spots, the inner ones of which correspond to these two spacings. Measurement of the R values, R_1 and R_2, of these inner diffraction spots are then used to give an accurate value of the camera constant,

$$\lambda L = R_1 d_1 = R_2 d_2.$$

5.2.3b *Polystyrene latex spheres*

A number of authors have examined both ordered and random arrays of polystyrene latex spheres using low angle electron diffraction techniques (Ferrier and Murray 1966; Smart and Burge 1965; Yeh and Geil 1967;

Glaeser and Thomas 1969). Suspensions of polystyrene spheres are available (see List of suppliers) in a range of sizes from 88 nm (800 Å) to about 1 μm diameter.

An ordered array of latex spheres, most commonly used as a calibration specimen, is prepared by placing a drop of a dilute suspension of the spheres in water on to a clean smooth substrate, preferably freshly-cleaved mica, and allowing it to air-dry. The particles gather together into an hexagonal array under the receding meniscus and when dry are coated with a thin film of evaporated carbon. The carbon film containing the latex spheres is floated off the mica substrate on a bath of water and collected on specimen grids. Alternatively a drop of the dilute suspension may be carefully placed onto a carbon or formvar-coated grid from a fine pipette, or the suspension may be sprayed over the coated grids until they are well wetted, and then allowed to dry as before. The latter technique is preferred if there is any tendency for the support films to shatter on contact with a drop of the suspension. The latex spheres will be observed packed hexagonally in rather small domains and some searching and experimentation with the preparation conditions may be necessary in order to achieve a satisfactory specimen.

The *d*-spacing of the hexagonal array is then simply the diameter of the spheres quoted for the particular batch of latex particles, and can be used along with the values of R measured from the diffraction pattern to determine the camera length λL, under the prevailing low angle diffraction conditions.

5.3 High resolution diffraction

5.3.1 INSTRUMENTAL DESIGN CONSIDERATIONS

The angular resolution, D/L, obtainable in the selected area, four-lens and five-lens modes, has been given as $\alpha_{min} = 4 \times 10^{-6}$ rad. This value is the highest resolution attainable in these modes of operation, assuming an infinitely small illumination aperture angle at the specimen, a condition only approached when both the C_1 and C_2 lenses are fully overfocused. Under these conditions the illumination intensity at the specimen is extremely low, and photographic exposure times are very long. Using normal illumination conditions, with the C_2 lens far less overfocused, the attainable resolution in these diffraction modes is typically, $\alpha_{typ} = 4 \times 10^{-5}$ rad.

In addition, the angular resolution in the diffraction pattern in these modes of operation may be further reduced by the effect of chromatic aberration of the objective lens, since there is a wide spread of electron energies in the beam after it has passed through the specimen. Also, possible instabilities in the projector lens currents may lead to blurring of the diffraction pattern.

To overcome these difficulties, in order to form a diffraction pattern of very high resolution, the technique of high resolution diffraction has been devised. In this mode of operation the specimen is mounted in a special stage below the final projector lens, and at a distance typically 250–300 mm from the photographic plate. With the specimen in this position, all of the electromagnetic lenses may be used, as required, to focus the electron beam to a very fine spot, thereby reducing the illumination aperture angle. In addition, since there are now no lenses between the specimen and the photographic plate, chromatic aberration and lens instabilities no longer have any serious effect.

In practice, illuminating conditions involving the use of C_1, the objective lens and the first projector lens P_1 are usually employed, when the highest angular resolution is required, giving $\alpha_{min} = 4 \times 10^{-7}$ rad. Other modes of high resolution operation are described in § 5.3.2.

A further advantage of the high resolution mode of diffraction is the constancy of the camera length, L, which is now the fixed physical distance between the specimen and the photographic plate (see § 5.3.3).

5.3.2 MODES OF OPERATION OF THE MICROSCOPE

The specimen is mounted in a special manipulator, generally known as a High Resolution Diffraction Stage, and inserted in the microscope through an airlock mechanism below the final projector lens. A typical stage and its mounting in the microscope are shown in Figs. 5.5 and 5.6. The specimen is usually held in a goniometer platform allowing full rotation and tilt of the specimen, as well as X and Y shift for area selection.

The illumination conditions can now be controlled by any combination of the condenser, objective and projector lenses, although certain combinations are preferred, as indicated below. The area of the specimen illuminated, and therefore the area from which the diffraction pattern is taken, is determined by the size of the aperture used in the selected area aperture plane and is usually about 0.5 × the diameter of the area-selecting aperture, i.e. a 50 μm area is selected by a 100 μm area-selecting aperture. When the illumination is focused to a point just above the specimen plane by adjusting the final lens in the particular combination of lenses being used, a shadow image of the specimen is formed at low magnification, from which a 50 μm area can be selected with reasonable accuracy. Having positioned the specimen, using the X and Y shift controls, so that the illumination is falling

Fig. 5.5. The AEI High Resolution Diffraction Stage.

Fig. 5.6. The High Resolution Diffraction Stage mounted in its position on the electron microscope below the final projector lens.

on the chosen area, the diffraction pattern is now formed by reducing the strength of the final lens until a focused central spot is seen on the viewing screen; other diffraction spots or rings will also appear, if the specimen is crystalline.

The particular illuminating conditions which should be used depend upon the degree of resolution required in the diffraction pattern, remembering that the greater the resolution required, the smaller the illuminating spot must be, leading to lower intensity in the diffraction pattern, and so to a longer photographic exposure. It must also be realized that the angular resolution attainable in the photographic plate is limited by scattering in the emulsion to approximately 10^{-5} rad.

The lens combinations generally considered are shown in Fig. 5.7. Lens system B, with C_1 as the source reducing lens and P_1 as the focusing lens, is generally preferred, giving a resolution comparable with that of the photo-

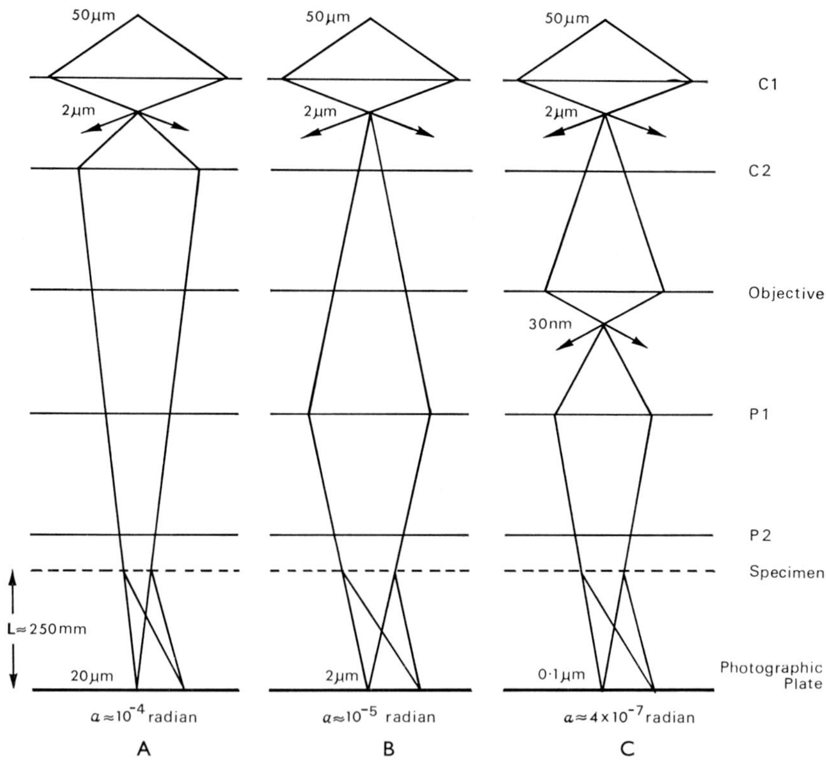

Fig. 5.7. Ray diagrams for the three standard modes of operation for high resolution electron diffraction.

graphic plate and a reasonable exposure time. If necessary the resolution can be improved considerably by using lens system C, with the C_1 and the objective lens as source reducing lenses, and P_1 as the focusing lens. However, much of the increase in resolution is likely to be lost due to scattering in the photographic emulsion, and also longer exposure times will be necessary. When only spot and ring positions are required, rather than high resolution, lens system A can be used, with C_1 and C_2 as the source reducing and focusing lenses respectively. Shorter exposure times are needed in this mode of operation, which may be useful for beam-sensitive materials.

5.3.3 CALIBRATION OF THE CAMERA CONSTANT, λL

As there are no longer any lenses between the specimen and the photographic plate, the camera length, L, is simply the physical distance between the specimen and the emulsion of the photographic plate. If the high resolution diffraction stage is accurately located on dowells in the microscope column airlock mechanism and the photographic plate is brought to the exposure position in the camera against well defined stops, the camera length will be accurately reproducible. The value of λL may be determined by using a standard thin film specimen of known d-spacings, as described in § 5.1.3.

Small changes in λL may occur when the specimen is tilted to a high angle but any inaccuracies introduced will be less than 0.2%. If necessary, a calibration chart may be drawn up for the value of λL for each setting of the diffraction stage controls, using a standard specimen. A very accurate value of λL is then quickly determined for an unknown specimen by noting the readings on the stage controls and referring to the calibration chart.

The camera length is usually fairly short, approximately 250 mm, in the high resolution diffraction mode and the diffraction beam stop should be used to prevent the inner diffraction rings and spots being lost in the diffuse scattering around the central spot due to dispersion in the photographic emulsion. Great care should be taken not to overexpose the photographic plate if the highest resolution is required in the diffraction pattern.

5.4 Reflection electron diffraction

A diffraction pattern may be obtained from the surface of a solid specimen, by mounting the specimen at a glancing angle to the incident electron beam in the microscope. Although the specimen is in a solid form, the result is still a transmission diffraction pattern as the electron beam penetrates the specimen to a depth of only a few nanometres, and the diffraction pattern

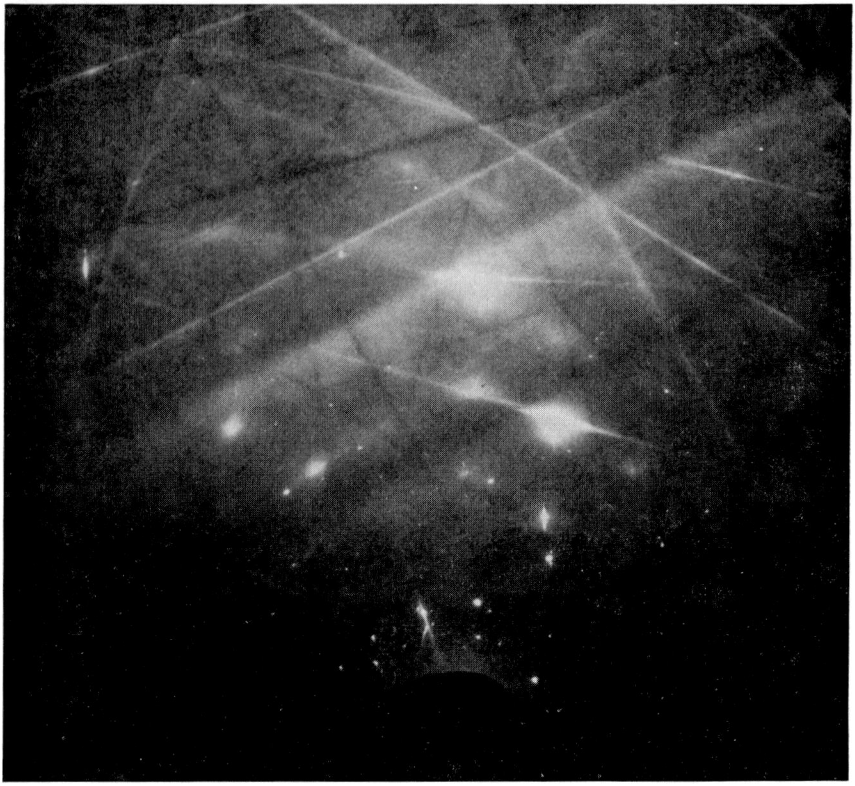

Fig. 5.8. Reflection electron diffraction pattern taken from copper oxide platelets formed on the surface of a copper specimen on heating to 900° C in a reflection/high resolution diffraction stage.

arises from this region. In this way the surface of a solid specimen or a thin deposited film or oxide layer may be examined. Monatomic layers of deposited films have been detected by reflection electron diffraction under carefully controlled conditions (Newman and Pashley 1955). The specimen surface should be fairly smooth and uncontaminated in order to form a good diffraction pattern (Fig. 5.8).

5.4.1 INSTRUMENTATION AND MODES OF OPERATION

A special manipulator is required to hold the specimen close to the optic axis of the microscope, with controls for X and Y shift of the specimen and for tilt about a horizontal axis, to adjust the glancing angle of the electron beam on the specimen surface. Two separate types of specimen stage have been used for this purpose.

5.4.1a *High resolution diffraction/reflection stage*

The standard high resolution diffraction stage, described in § 5.3.2, can usually be used in a reflection diffraction mode of operation by mounting the specimen on a stub which fits into the standard holder, and tilting the specimen stage until the specimen surface is almost parallel to the electron beam. A variety of specimen holders is generally supplied to allow specimens of many different shapes and sizes to be held in the stage. If the specimen must be glued to the stub to make a satisfactory mechanical contact (cellulose acetate adhesive or 10% collodion in amyl acetate may be used successfully), it is important to establish good electrical contact also, or else the specimen surface will become charged by the electron beam and distort the diffraction pattern. This is easily done by painting on a strip of colloidal graphite, Aquadag (see List of suppliers), to bridge the gap between the specimen and the support stub.

The specimen holder or mounting stub is clamped in the diffraction stage, care being taken to position the specimen so that the surface to be examined lies on the tilt axis of the stage. This is necessary to ensure that when the specimen is tilted, to alter the glancing angle of the electron beam on the specimen surface, no horizontal shift of the specimen occurs, which would alter its position in relation to the optic axis. The tilt controls should be set so that the specimen surface is tilted slightly upwards from the vertical, making a glancing angle of only a few degrees.

The diffraction stage is inserted through the airlock into the microscope in its usual position below the final projector lens. The electron beam may be focused on to the screen by any combination of the electron lenses in the microscope, although one of the combinations of lenses used in the high resolution diffraction mode, shown in Fig. 5.7, is usually adopted, depending upon whether or not high resolution is the predominant requirement, as before.

If no illumination spot appears on the viewing screen it is likely that the electron beam is being obstructed by the specimen, which should then be withdrawn using the shift controls until the spot appears. The specimen should then be returned towards the electron beam, while simultaneously making small adjustments to the glancing angle, until a reflection diffraction pattern is formed on the screen. The quality of the diffraction pattern is very dependent upon the nature of the specimen surface and the glancing angle, and it is recommended that a preliminary experiment is conducted on a freshly deposited polycrystalline gold film in order to gain experience and confidence in the technique.

A rough area selection is again possible by focusing the illumination just above the specimen and casting a shadow image on the viewing screen, as in § 5.3.2.

5.4.1b *Reflection imaging and diffraction stage*

Reflection specimens may also be mounted in a special stage in the normal specimen chamber of some microscopes. In this position the microscope imaging lenses may be used to form both a reflection image and a reflection diffraction pattern of the specimen surface. However, reflection imaging of specimen surfaces has now been almost entirely replaced by shadow/replication techniques and scanning electron microscopy, and stages for electron diffraction of massive specimens in the normal specimen chamber are no longer generally available.

Stages for electron diffraction analysis of normal 3.05 mm diameter thin film specimens are available for certain instruments. The microscope is operated in the normal diffraction mode with the objective lens at substantially its normal focal length. A reflection image is not available with this specimen holder as insufficient tilt of the illumination is possible to show the surface topography in adequate relief.

Older instruments, for which a reflection imaging and diffraction stage is available, will accept a relatively massive specimen. The objective lens pole-pieces must then be removed, where this is possible, or the specimen must be mounted above the objective lens, in order that the necessary specimen movements can be achieved.

The objective lens is now operated at a long focal length in order to view the diffracting surface of the specimen. The camera length, L, is subject to considerable variation because of the unusually large depth of the specimen in the direction of the electron beam. The diffraction spots are therefore considerably broadened and the angular resolution is very poor in this mode, although it does have the advantage that an image of the diffracting surface is obtainable and rough area selection for diffraction analysis is possible.

5.4.2 REFLECTION DIFFRACTION FROM NON-CONDUCTING SPECIMEN SURFACES

A specimen surface which is non-conducting will charge up under the electron beam, and this charge will grossly distort the electron diffraction pattern. In order to avoid this problem specimen surfaces, such as oxide layers, refractory oxide specimens or other insulating materials, must be discharged by a shower of electrons accelerated at low voltage by an ancillary electron discharging gun. Such discharging guns are generally available as an accessory with a diffraction/reflection stage. The gun is mounted on the stage, directed at the specimen surface, and the filament current of the electron gun increased until the charging effect disappears from the shadow image or the diffraction pattern. The charging effect may reappear if the specimen beam current, the condenser lens currents or the specimen position are altered significantly, but can be quickly removed again by adjustment of the filament current of the discharging gun.

5.4.3 INTERPRETATION OF THE REFLECTION DIFFRACTION PATTERN

As the reflection diffraction pattern is essentially a transmission diffraction pattern through a small volume of material on the surface of the reflection specimen, the interpretation is fundamentally no different from that of a transmission electron diffraction pattern described previously. Some differences in the patterns however will be noticed.

(a) The diffraction spots are likely to be broad, as previously mentioned, because the camera length may have a wide range of values, rather than a single well defined value as in transmission diffraction. Calibration of the camera length using known standards is thus bound to be equally inaccurate.

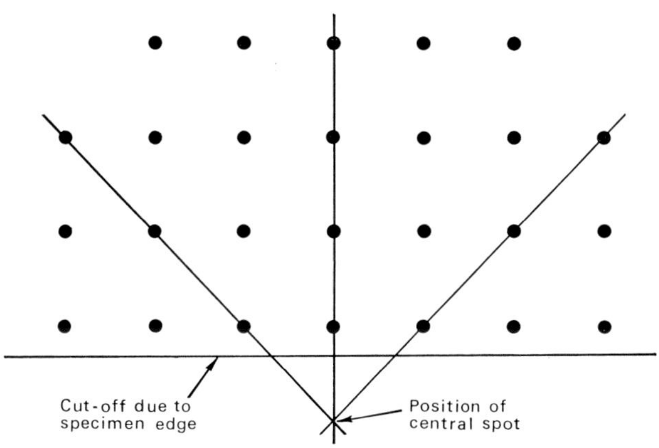

Fig. 5.9. Positioning the 'central spot' of a reflection diffraction pattern by extrapolating the major zones of diffraction spots.

(b) Only one 180° segment of the diffraction pattern will be observed, the other segment being obstructed by the edge of the specimen, and on occasions even the central spot of the diffraction pattern will not appear. This makes measurement of the R values rather difficult unless the position of the 'central spot' can be determined by the intersection of lines extrapolated from the major zones of diffraction spots, as in Fig. 5.9.

(c) In addition to the broadening of the diffraction spots, many dark and bright lines will appear in the reflection diffraction pattern, known as Kikuchi lines. These lines appear, both in reflection and transmission diffraction patterns, whenever the electron beam is penetrating a thick specimen, as is very often the case in reflection diffraction experiments. The formation and interpretation of Kikuchi lines will be discussed in some detail in § 7.1.

5.5 Scanning electron diffraction

The conventional photographic procedure for recording diffraction patterns in the electron microscope suffers from four major disadvantages.

(i) Only a rather inaccurate measure of the intensities of the diffraction spots and rings is available from a densitometer trace across the photographic plate. While relative intensities may be obtained with some accuracy, absolute measurements are very dependent upon the exposure and developement procedure and are generally insufficiently reliable.

(ii) Both the inelastically and elastically scattered electrons are recorded simultaneously in the diffraction pattern, precluding any structural analysis based upon the intensities of the diffraction spots.

(iii) The procedure is fairly lengthy, and not suitable for recording dynamic changes occurring in the diffraction pattern, particularly if immediate interpretation and reaction to any change in the diffraction pattern is necessary.

(iv) The data are not obtained in a form suitable for rapid numerical manipulation and analysis.

A technique which overcomes these deficiences, with the particular advantage that energy-loss electrons may be eliminated from the diffraction pattern by energy filtering, is scanning electron diffraction.

5.5.1 INSTRUMENTATION

The equipment consists essentially of a raster scanning system, an energy filter, an electron detector and a display readout system. Although the energy filter may be either magnetic or electrostatic in design, the electrostatic filter is generally preferred for simplicity of control and readout. A scanning electron diffraction system is shown schematically in Fig. 5.10, and electron microscope attachments for scanning electron diffraction analysis, SEDA, are now commercially available (Barden and Craig-Gray 1970).

The scanning system consists of two sets of coils, X and Y, mounted on a torroidal former, inserted between the P_1 and P_2 lenses of the microscope. When the scan coils are excited by the appropriate currents from the scan generator, the diffraction pattern, formed in any one of the standard modes described in §§ 5.1, 5.2 and 5.4.1b, may be swept over a defining aperture mounted in the viewing screen or in the base of the microscope camera. The defining aperture may be as small as 25 μm, thereby selecting very fine elements of the diffraction pattern.

The electrons forming the diffraction pattern which are momentarily over the defining aperture, pass through and impinge on the energy filter. The electrostatic filter consists of two electrodes held at earth potential, between which is a third torroidal electrode held at a high voltage close to that of the electron gun of the microscope column. The potential of this central electrode

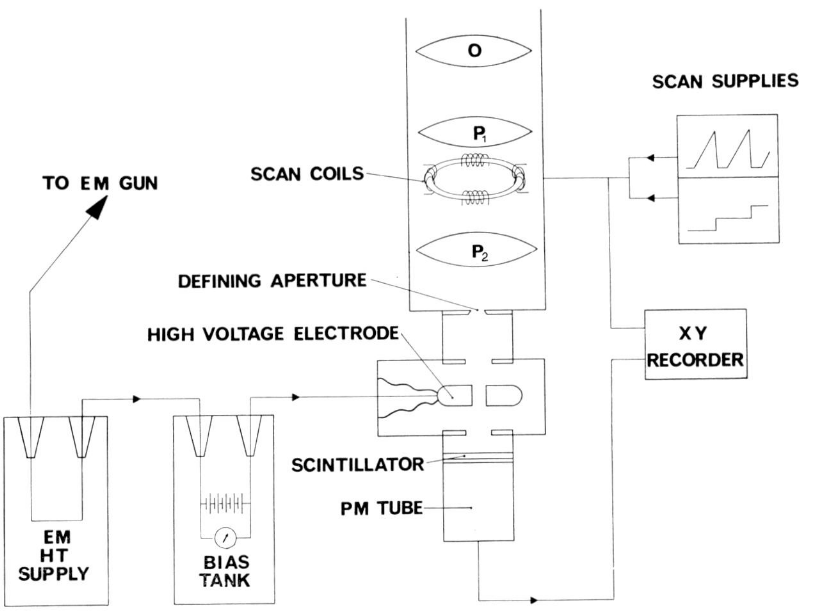

Fig. 5.10. Schematic diagram of a scanning electron diffraction attachment, SEDA, with energy filtering, for an electron microscope.

is obtained from an extra output from the electron microscope high tension supply to the normal electron gun. This output is taken to an oil-filled bias tank, where a metered battery bias, which may be varied from −30 V to +300 V, is added to the potential before it is applied to the high voltage electrode of the filter. In order to prevent discharge of the high voltage electrode, the filter chamber is made of stainless steel and pumped to a high vacuum with its own vacuum system.

Below the energy filter is a lead-glass mounted phosphor scintillator, followed by a photomultiplier tube (PM). The scintillations detected by the photomultiplier are fed to the Y-axis of an XY recorder, the recorder's X-axis being swept in synchrony with the sawtooth curent applied to the X scan coils.

5.5.2 OPERATION

The diffraction pattern is formed on the viewing screen in the usual way and the bias on the energy filter set to pass only those electrons which have lost no energy or less than a few volts of energy in passing through the specimen,

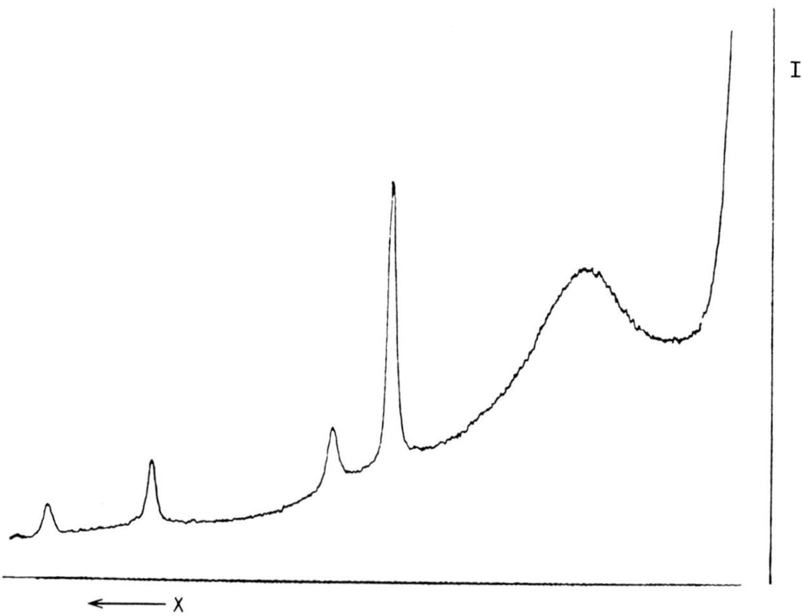

Fig. 5.11. A single intensity (I) scan across the filtered polycrystalline diffraction pattern of a 4nm thick aluminium film on a collodion substrate: without filtering, the first broad peak could not be detected. (From Denbigh and Grigson 1965.)

for instance +5V. Under these conditions, the inelastically scattered electrons, which have lost energy and give rise to the diffuse scattering around the central diffraction spot, will be eliminated, and more accurate values of the intensities of the elastically scattered electrons in the diffraction spots close to the central spot will be recorded. This is particularly important for low angle diffraction patterns from materials with only pseudo-crystalline arrays. If the X scan coils are energized simultaneously with the recording system, and a single traverse of the diffraction pattern across the defining aperture made, a display of the intensities and the spatial distribution of elastically scattered electrons across the diffraction pattern will be obtained, as Fig. 5.11 (Grigson 1962; Denbigh and Grigson 1965). If, in addition, a staircase current form is applied to the Y scan coils and to the Y axis of the chart recorder, a record of the intensity distribution throughout the whole diffraction pattern may be obtained, of the type shown in Fig. 5.12 (Grigson and Tillett 1968).

Alternatively, the filter may be set to pass a certain energy loss band rather than the no-loss electrons, and considerable further electronic and numerical manipulation of the output signal is possible which is outside the scope of this text (see Denbigh and Grigson 1965; Grigson and Tillett 1968).

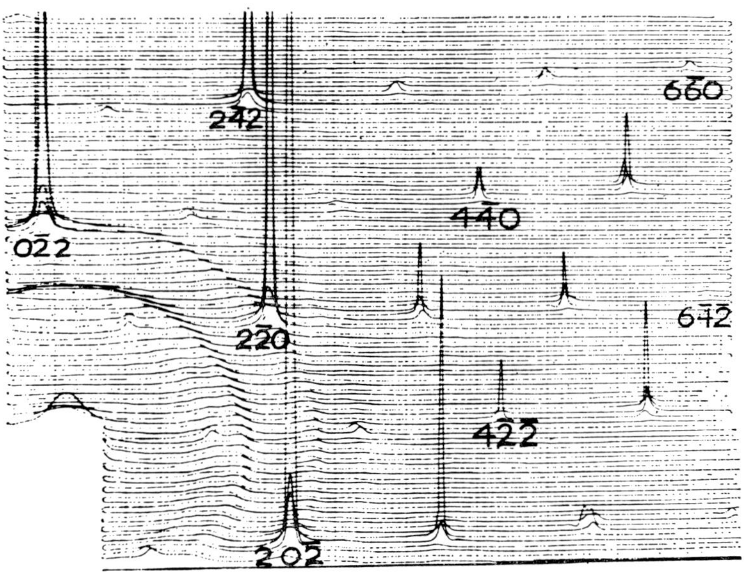

Fig. 5.12. 'Three-dimensional' recording of a filtered scanning electron diffraction pattern of a single crystal specimen of gold. (From Grigson and Tillett 1968.)

By operating the microscope in the transmission imaging mode rather than the diffraction mode, and placing a feature of interest in the image over the defining aperture, measurement of the plasmon energy losses, or X-ray losses, in that particular area of the specimen is possible. Studies of the effect of alloying on the plasmon energy losses in certain alloy systems may be made, and used for the micro-analysis of very small specimen areas, particularly of aluminum alloys.

5.5.3 APPLICATIONS OF SCANNING ELECTRON DIFFRACTION

The scanning system for recording diffraction patterns after energy filtering is used in

(a) determinations of the structure of 'amorphous' thin films, that is, the degree of order, the nearest neighbour distances and the coordination number, (the number of nearest neighbours),

(b) the study of dynamic transformations initiated by such processes as specimen heating, cooling or straining,

(c) the detection of low angle scattering phenomena from thin specimens of relatively amorphous materials, made possible by the filtering of the inelastically scattered electrons around the undeviated diffraction spot, and

(d) the study of many fundamental electron scattering problems.

REFERENCES

Agar, A. W. (1960), Accuracy of selected-area microdiffraction in the electron microscope, Brit. J. Appl. Phys. *11*, 185.

Armitage, W. K. and A. MacConaill (1964), Variations in diffraction camera factor for the Seimens Elmiskop 1, J. Sci. Instr. *41*, 401

Barden, L. F. and J. Craig-Gray (1970), Scanning electron diffraction attachment with energy filter, Proc. 7th Int. Cong. on Electron Microscopy, Grenoble *1*, 181.

Bassett, G. A. and A. Keller, (1964), Low angle scattering in the electron microscope: application to polymers, Phil. Mag. *9*, 817.

Curtis, G. H., R. P. Ferrier and R. T. Murray (1967), An improved technique for ultra-high resolution electron diffraction, J. Sci. Instr. *44*, 867.

Denbich, P. N. and C. W. B. Grigson (1965), Scanning electron diffraction with energy analysis, J. Sci. Instr. *42*, 305.

Ferrier, R. P. and R. T. Murray (1966), Low angle electron diffraction. J. Roy Microscop. Soc. *85*, 323.

Ferrier, R. P. (1967), The technique and applications of low angle electron diffraction, Bull. Soc. Franç. Mineral. Crist. *XC*, 464.

Glaeser, R. M. and G. Thomas (1969), Application of electron diffraction to biological electron microscopy, Biophys. J. *9*, 1073.

Grigson, C. W. B. (1962), On scanning electron diffraction, J. Electron. Control *12*, 209.

Grigson, C. W. B. and P. I. Tillett (1968), On scanning electron diffraction II, Int. J. Electron *24*, 101.

Hambling, P. G. (1953), The lattice constant and expansion coefficients of some halides, Acta Cryst. *6*, 98.
Newman, R. C. and D. W. Pashley (1955), The sensitivity of electron diffraction as a means of detecting thin surface films: I, Phil. Mag. *46*, 927.
Phillips, R. (1960), Selected-area diffraction in the electron microscope, Brit. J. Appl. Phys. *11*, 504.
Reicke, W. D. (1961), The exactness of agreement between selected and diffracting areas in Le Poole's selected-area diffraction technique, Optik *18*, 278; Distortion and resolution in diffraction patterns obtained by Le Poole's Technique, Optik *18*, 373.
Rickards, G. K. and G. K. Williamson (1964), Instrumental constants in Philips EM 200 microscope, J. Sci. Instr. *41*, 174.
Smart, J. and R. E. Burge (1965), Low angle electron diffraction pattern of assemblies of spheres and viruses, Nature *205*, 1297.
Wrigley, M. G. (1968), The lattice spacing of crystalline catalase as an internal standard of length in electron microscopy, J. Ultrastruct. Res. *24*, 454.
Yeh, G. S. Y. and P. H. Geil (1967), Selected-area small-angle electron diffraction, J. Mater. Sci. *2*, 457.

Chapter 6

The applications of electron diffraction analysis

Whenever the specimen, or the selected specimen area under examination in the electron microscope, is crystalline in character electron diffraction analysis will provide much information which is not so specifically and accurately obtained by other means. The basic information that can be obtained from a single crystal diffraction pattern is

(i) the ORIENTATION of the planes of atoms or molecules with respect to the electron beam and to the image of the specimen,

(ii) the orientation of the planes of atoms or molecules with respect to each other, that is, the crystallographic STRUCTURE of the specimen area, and

(iii) the *d*-spacings between the planes, from which it may be possible, in conjunction with (ii) to determine the chemical COMPOSITION of the specimen area.

In general terms, the information in (i) will be of most interest to the metallurgist, as it enables him to determine the orientation of defects, precipitates and inclusions with respect to the matrix, and the nature of the strain fields surrounding these features. The determination of the specimen structure (ii), is the role of the crystallographer, who may be examining metallurgical, biological or other materials, for correlation with other analyses in any of these disciplines. The compositional information (iii), will be of interest to the investigator with an unknown material in his specimen, be it, for example, a foreign body in a section of lung tissue, an unknown inclusion in a refractory composite or an undetermined precipitate in a metal alloy.

6.1 Orientation determinations

6.1.1 ORIENTATION RELATIONSHIPS BETWEEN DIFFERENT AREAS OF THE SPECIMEN

The orientation relationship between

(a) two crystalline inclusions in an amorphous matrix,

(b) two adjacent crystal grains in a polycrystalline specimen, separated by a grain boundary or twin boundary

(c) a crystalline precipitate or inclusion and the crystalline matrix in which it has grown or has been embedded, are typical examples of information which is readily achieved by forming a diffraction pattern from each of the two separate areas to be related.

It is important to ensure that the diffraction pattern has been obtained accurately from the areas of interest, thus effectively limiting this type of study to specimens with sufficiently small lattice spacings that a useful selected area diffraction, SAD, pattern can be formed, and to specimen areas not less than 1 μm in diameter. Thus it may be very difficult to study the orientation relationship between a 50 nm diameter precipitate and the alloy matrix, or between two small precipitates less than 1 μm apart. For these specimens, the diffraction information will be superimposed in a single diffraction pattern, and it may be difficult to determine which spots come from which crystal, and thus determine the orientation relationship between the two crystals.

It is often possible to assign each diffraction spot in a composite diffraction pattern to a particular crystal, by forming a 'dark field' image using each diffraction spot in turn, when the crystal from which a particular spot originates will appear bright in the final image, while the other crystals and the matrix will appear dark. A 'dark field' image may be formed either by displacing the objective aperture to surround a diffraction spot rather than the central spot in the diffraction pattern, or preferably by tilting the incident illumination to bring a diffraction spot into the position in the diffraction pattern normally occupied by the central spot, and then switching back to the imaging mode of operation. In this way it may be possible to seperate and analyse two superimposed diffraction patterns, and thus determine their orientation relationship.

However, it is generally preferable to form individual diffraction patterns from the two separate areas of the specimen where possible, and then to index the patterns and determine the zone axis of each pattern, as indicated in §§ 4.2 and 4.3. The orientation relationship between the two areas is then

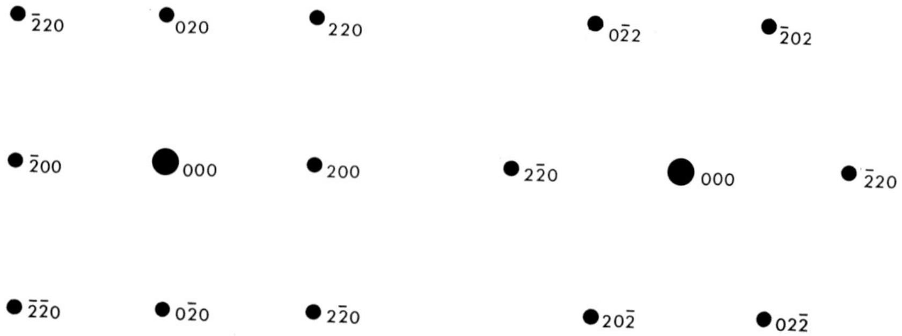

Crystal A, [001] zone axis Crystal B, [111] zone axis

Fig. 6.1. Determination of the orientation relationship between two crystals of an f.c.c. material by analysis of their diffraction patterns.

defined by stating the two determined zone axes as parallel to each other, and the indices of the direction in the plane of one of the diffraction patterns which is parallel to a selected direction in the plane of the other diffraction pattern, as in Fig. 6.1. In this example, crystals A and B have f.c.c. structures, and the orientation relationship between the two crystals may be quoted as

$$[001]_A \| [111]_B \quad \text{and} \quad [200]_A \| [\bar{2}20]_B.$$

Determination of the orientation relationship between two crystals using only the diffraction spots of the patterns will give a result accurate to only a few degrees. This is because the reciprocal lattice spots are extended in the direction of thinness of the specimen (§ 4.7.2), and so a thin crystal may be tilted through a small angle and still maintain the same array of diffraction spots, even though the intensities may alter. Various methods of 'weighting' the intensities to determine the angle of tilt from the symmetry position, in order to achieve more accurate determination of crystal orientation, have been described (Laird et al. 1966; Shenin 1965), although their accuracy is suspect if the specimen is at all curved or buckled. The preferred technique for accurate orientation determination makes use of the Kikuchi lines (§ 7.1), which often appear in the diffraction pattern in addition to the diffraction spots. As the crystal is tilted, the Kikuchi lines sweep across the diffraction pattern as though they were firmly attached to the specimen, and so their displacement in the diffraction pattern from the associated diffraction spots may be used as an accurate measure of the crystal orientation with respect to the electron beam (Otte et al. 1964; Von Heimendahl et al. 1964; Hirsch et al. 1965; Shenin and Cahn 1965; Andrews et al. 1971).Ryder and Pitsch (1968) have shown that an accuracy of $\pm 0.1\%$ can be achieved in orientation determinations by analysing the Kikuchi line patterns.

6.1.2 ORIENTATION OF IMAGE PHENOMENA WITH RESPECT TO THE CRYSTAL LATTICE

Frequently it is necessary to determine the direction in which a particular phenomenon is lying in the crystalline specimen, such as a dislocation or stacking fault in a metal specimen, or the edge of an isolated crystallite lying on a support film (as used in the Image Rotation Calibration technique, described in § 6.1.4), or a boundary or discontinuity in a biological crystal. This may be done by taking a micrograph and a diffraction pattern of the specimen area containing the phenomenon of interest, indexing the diffraction pattern, as in §§ 4.2–4.4, and finding the direction in the diffraction pattern or appropriate stereographic projection (see Johari and Thomas 1969), which is parallel to the phenomenon of interest in the micrograph. This is the direction of the projection of the phenomenon into the plane normal to the electron beam. However, this may not necessarily be the actual direction of the phenomenon in the crystal lattice, as the specimen may have a finite depth, and the phenomenon, perhaps a dislocation line, AB, may lie through the specimen at an angle to the two surfaces (Fig. 6.2). For a full determination of the line direction in this case, two micrographs and diffraction patterns should be taken with the specimen tilted at widely different angles, and the actual direction in the crystal determined from the two projected directions. This type of 'trace analysis' is described in general terms by Barrett (1952), Phillips (1955) and Wood (1963), and with particular reference to electron micrographs and diffraction patterns by Thomas (1962), Hirsch et al. (1965) and Andrews et al. (1971).

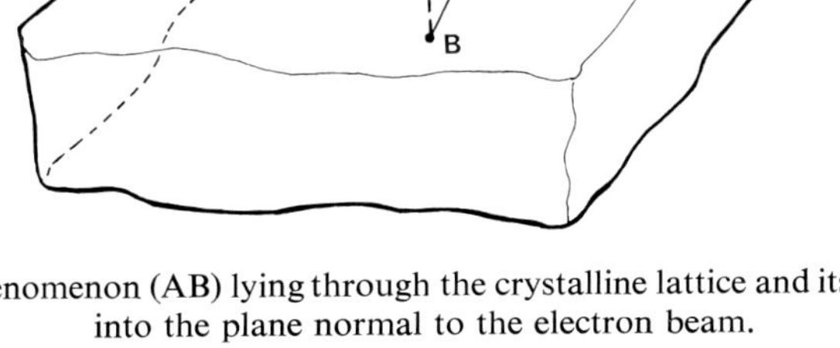

Fig. 6.2. A phenomenon (AB) lying through the crystalline lattice and its projection (AC) into the plane normal to the electron beam.

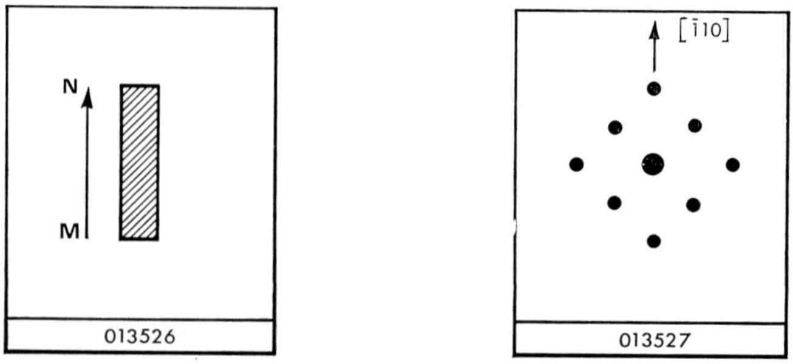

Fig. 6.3. Incorrect correlation of the electron microscope image and the diffraction pattern, as no allowance has been made for electron-optical rotation.

One further correction must be made before a direct correlation of a direction in the diffraction pattern and the direction in the micrograph can be made. It cannot be assumed that if the micrograph and the diffraction pattern are laid side-by-side on the desk, in the same orientation, as in Fig. 6.3, that parallel directions in the two can be immediately correlated. That is, it cannot be assumed that the direction of the edge of the particle, MN, can be immediately read off as $[\bar{1}10]$. This is because the image and the diffraction pattern are not straight projections of the specimen in the electron microscope on to the viewing screen, as they are in a light microscope, but are images formed by electrons.

The electron images, in addition to being magnified by the electro-magnetic lenses, also undergo a rotation. Since the P_1 lens must be operated at different strengths in order to focus first the image and then the diffraction pattern on to the viewing screen or photographic plate, a marked difference in the 'image' rotations on passing through the P_1 lens will occur in the two cases. The further rotation which the first image undergoes before reaching the viewing screen or photographic plate, depends upon the total magnification of the final image, and the full rotation of the diffraction pattern depends upon the diffraction camera length, since both these two parameters depend upon the settings of the projector lenses. So the rotation difference between the image and the diffraction pattern, the amount by which the one must be rotated with respect to the other in order to correlate parallel directions in the two, varies with magnification and camera length.

A set of curves relating rotation difference, magnification and camera length may be calculated from information available from the manufacturers, but is more often determined experimentally using a molybdenum trioxide crystal test specimen, which is known to have crystal edges in certain well-defined directions. By superimposing an image of the crystal on top of the diffraction pattern by a double exposure on a single plate, experimental measurement of the rotation difference may be made at each magnification and each camera length. This calibration technique is described in full in § 6.1.4.

In addition, a further difference in rotation of 180° must be taken into account, since the image is inverted through 180° between the object and the first image, yet no such inversion occurs to the diffraction pattern (Fig. 3.4). So 180° must be added to the rotation difference between the final image and the diffraction pattern (Groves and Whelan, 1962).

It should also be remembered that the photographic plate is always exposed to the electron beam with the emulsion facing upwards, for microscopes with the electron gun at the top, whereas the negatives are usually printed with the emulsion facing downwards. This inversion occurs for both the image and the diffraction pattern, but means that a clockwise rotation difference becomes an anti-clockwise one, and vice-versa. To obviate the need for this inversion, the negatives can of course be printed with the emulsion

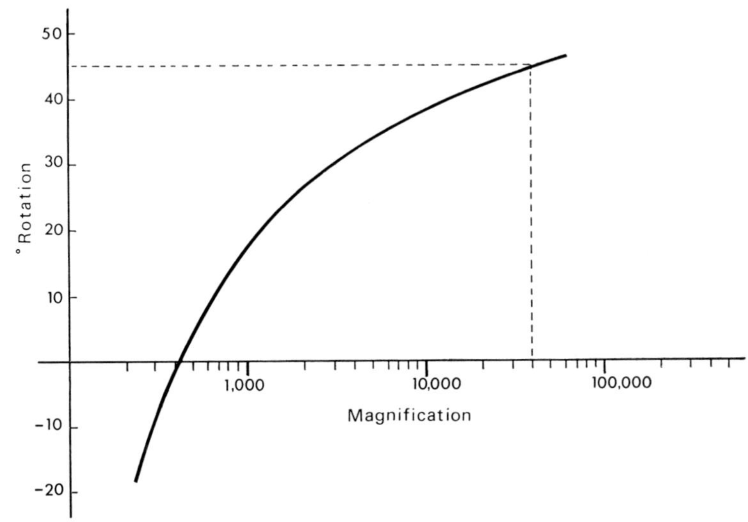

Fig. 6.4. SCHEMATIC rotation difference curve for change of manification only (assuming a fixed camera length)–individual rotation difference curves should be determined experimentally for each microscope before correlation of directions in the image and the diffraction pattern can be obtained.

facing upwards, accepting the small loss in image quality resulting from this procedure.

In order to demonstrate the use of such a rotation difference calibration curve, a schematic curve is illustrated in Fig. 6.4. Since the majority of electron microscopes have only a single SAD camera length, generally about 500 mm, a schematic rotation difference curve for change in magnification only is drawn. Now with reference to this curve, if the micrograph in Fig. 6.3 was recorded at 40,000 ×, it must be rotated in a clockwise sense through 45° before an immediate correlation between a direction in the micrograph may be made with the parallel direction in the diffraction pattern (Fig. 6.5).

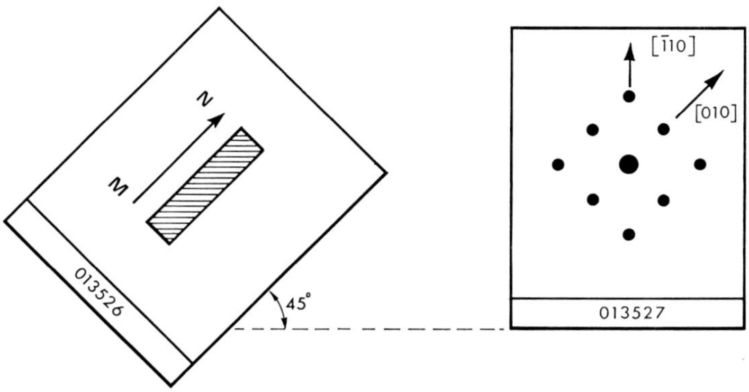

Fig. 6.5. Correlation of directions in the image and the diffraction pattern, after allowance has been made for electron-optical rotation.

Thus, the long edge of the crystal MN is actually a [010] direction, and not a [$\bar{1}$10] direction as might originally have been supposed. It is now a simple matter to correlate other directions in the micrograph with the parallel direction in the diffraction pattern, the correct rotation difference once having been established.

6.1.3 ELECTRON DIFFRACTION IN DEFECT CONTRAST ANALYSIS

Images of 'defect' phenomena in crystalline specimens, such as, dislocations, stacking faults, fine precipitates and voids, which are formed by diffraction contrast mechanisms, are very sensitive to specimen orientation. Images are formed of such phenomena when the stress fields surrounding them are sufficient to distort the lattice planes locally to an extent that they then lie at a Bragg angle to the incident electron beam, and thus cause the electrons striking that particular local area to be diffracted outside the objective aperture. In this way a dark line or area will appear on the photographic plate, from the region around the defect, as shown in Fig. 6.6. In order that the strains around this type of defect may be sufficient to tilt the lattice planes locally into a Bragg angle, the particular set of planes must be very close to a Bragg angle initially. It is in this initial orientation of the specimen to position a particular set of lattice planes close to a Bragg angle that electron diffraction information is very important.

Remember that any one particular set of lattice planes gives rise to a single spot in the diffraction pattern, and if the specimen is tilted until only this diffraction spot and the central spot appear in the diffraction pattern, the so-called 'two-beam' condition, then only this one set of lattice planes is reflecting the electrons outside the objective aperture. Under these conditions,

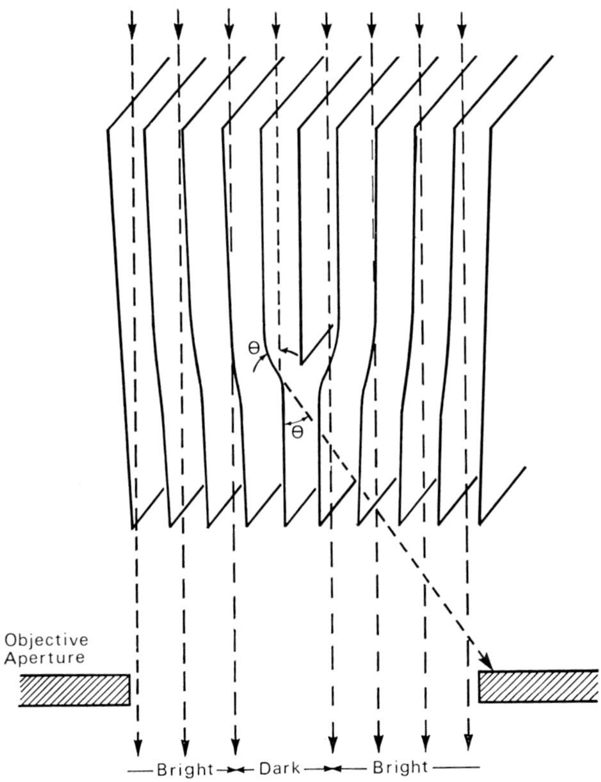

Fig. 6.6. Schematic representation of the formation of the diffraction contrast image from an edge dislocation in a crystalline lattice.

a perfectly flat specimen would give a totally dark image on the viewing screen. However, specimens are usually at least slightly curved, and dark bands then appear in the image where the lattice planes are exactly at a Bragg angle with light areas on either side where the planes are slightly deviated from the Bragg condition. It is in these light areas that lattice defects are generally observed in optimum contrast, as the strain surrounding the defects may be just sufficient to tilt the local lattice planes back into the Bragg condition.

If an image of the defect is observed, then the defect must be bending the particular set of planes out of their original plane, as in Fig. 6.7a, whereas, if the defect causes a distortion which only shifts the atoms within the plane, as in Fig. 6.7b, the electron beam would not detect any displacement, and

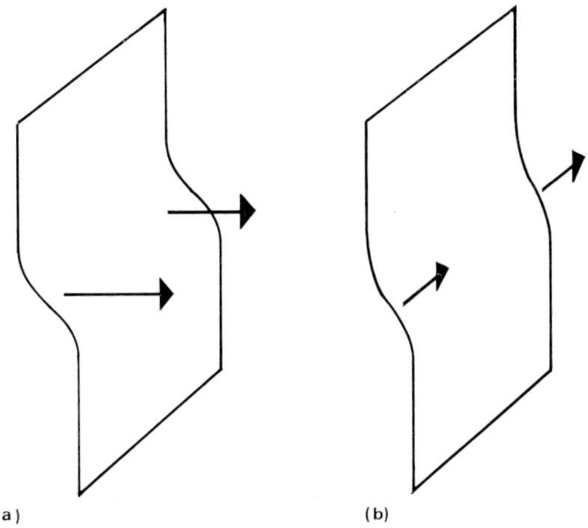

Fig. 6.7. Atomic displacements caused by lattice defects which a) distort a set of lattice planes out of their original plane, and b) only shift the atoms within the original plane.

an image is not observed – the defect is 'out-of-contrast'. It is possible to determine uniquely the direction of a linear displacement in the crystal lattice, such as the Burgers' vector of a dislocation in a metal foil, by finding two sets of planes in which the atoms are only shifted within the plane, that is, two different sets of reflecting conditions under which the dislocation is 'out-of-contrast'. The line of intersection of the two planes is then the direction of the displacement. To do this, successive two-beam diffraction patterns are formed of the same area of the specimen, containing the defect of interest, using different diffraction spots, and the image formed under each of these conditions until the orientations are found in which the defect is out-of-contrast. Experiments of this type require an accurate specimen tilting stage, which allows at least $\pm 30°$ of tilt about any axis in the horizontal plane.

Burgers' vector determination is one example of the type of contrast analysis experiment which may be performed in the electron microscope, which places great reliance upon the information obtained from the electron diffraction pattern formed from the specimen area of interest. In practice, much use is also made of the Kikuchi line patterns (§ 7.1) to determine the exact orientation of the specimen area. Many further examples of contrast analysis experiments are to found in texts by Thomas (1962) and Hirsch et al. (1965), and in articles by Hirsch et al. (1960), Steeds and Hirsch (1963), Gevers (1963), Ashby and Brown (1963), Silcock and Tunstall (1964), France and Loretto (1968) and Cockayne et al. (1971), to list but a few.

6.1.4 IMAGE ROTATION CALIBRATION

The extent of image rotation at the different magnifications of the microscope, and of rotation of the diffraction pattern at different camera lengths, with respect to the specimen, is usually determined experimentally by superimposing the image of the long edge of a molybdenum trioxide crystal on top of its diffraction pattern by double exposing a single photographic plate.

The molybdenum trioxide, MoO_3, test specimen is prepared from the standard crystalline material obtainable from the usual chemical suppliers (see List of suppliers). Place a few grams (a small spatula-full) of MoO_3 crystals in a 2.5 cm diameter 'boiling tube' and gently heat the end of the tube, held in an almost horizontal position, over a bunsen flame until the crystals begin to sublime, to drive off any volatile impurities. Allow the crystals to cool, and then slide into the open end of the tube a clean glass slide on which have been positioned a number of carbon-coated specimen grids. Close up the tube with a loosely-fitting bung, and reheat the MoO_3 crystals until a visible quantity is seen to sublime and recondense onto the glass slide. Allow to cool and remove the specimens from the tube.

The carbon support films should now be covered with large numbers of electron-transparent MoO_3 crystals with a well-defined shape – rectangular with the four corners cut off at 45°. The long edge of each crystal is known to be an [001] direction in the orthorhombic crystal lattice. This corresponds with the long direction in the rectangular array of spots in the diffraction pattern, for which $d_{001} = 0.3696$ nm.

Under certain diffraction conditions it is possible that some of the diffraction spots in the pattern may be very weak, giving the impression of a rectangular array of spots rotated through $\sim 45^\circ$ from the correct rectangular array. Thus it may be necessary to check that the spacing of the diffraction spots along the long direction of the rectangular array does correspond to a d-spacing of 0.3696 nm for the particular value of λL being used.

The procedure is then to form an image of the long edge of a large MoO_3 crystal at one magnification, record the image photographically, and then before returning the plate to the camera box form the diffraction pattern at one camera length and superimpose the diffraction pattern on the recorded image. The procedure should be repeated for the full range of magnifications at the same camera length, and then a second set of plates taken over the

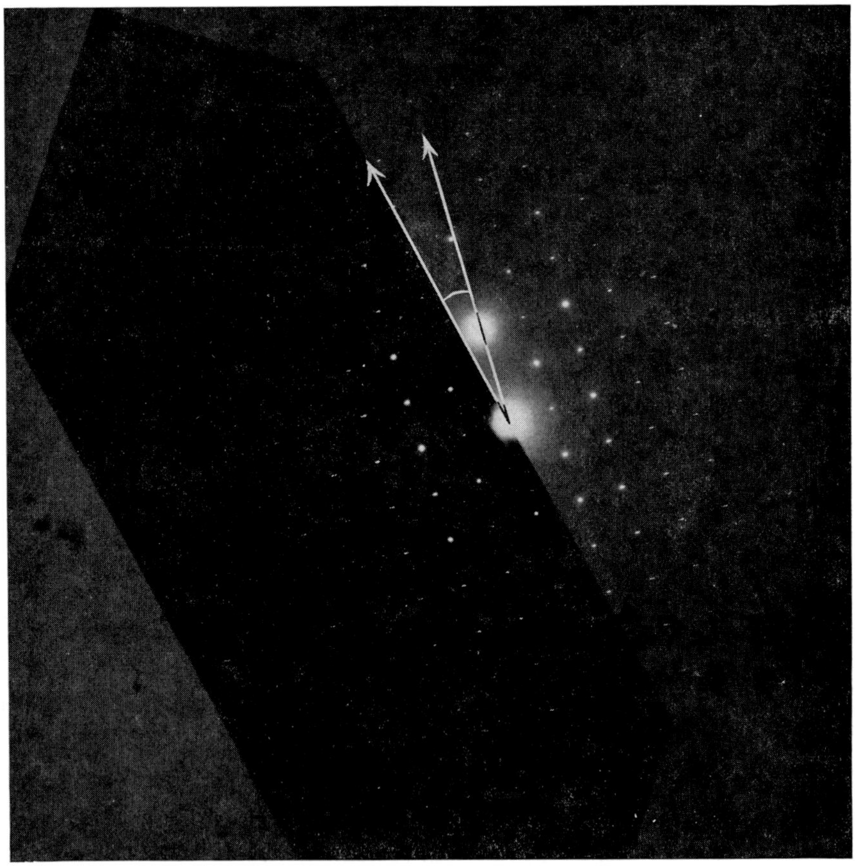

Fig. 6.8. Single crystal diffraction pattern of a molybdenum trioxide crystal superimposed on the image of the long edge of the crystal.

complete range of camera lengths if more than one camera length can be obtained, maintaining the image at a convenient magnification.

The rotation differences between the image at each magnification and the diffraction pattern at the one camera length are then easily measured, as in Fig. 6.8, and plotted on a chart as in Fig. 6.4, after addition of 180° to allow for the inversion of the image with respect to the diffraction pattern through the objective lens. The extra rotation factor to be added or subtracted from the rotation difference to allow for the rotation of the diffraction pattern with change of camera length can then be measured from the second set of plates and recorded in a table.

To check the sense of the rotation of the image with respect to the diffraction pattern, the diffraction pattern should be formed and then the strength of the first projector lens increased so that each of the diffraction spots expands into a small image of the crystal edge; further increase of the strength of the P_1 lens will then make the rotation of the image quite apparent. Remember to allow for any inversion of the photographic plates which may take place during printing of the negatives, as mentioned in § 6.1.2. Such an inversion will alter the sense of the rotation difference but not the magnitude.

6.2 Determination of specimen structure

It is not possible to lay down a rigorous procedure to follow in order to determine the crystal lattice of a specimen, as there are a large number of possible starting points depending upon the amount of corroborative information about the specimen that is available. The possibilities range from (a) a crystal whose elemental composition is well known, but which may crystallize in either one of two known structures, through to (z) a completely foreign specimen, about which nothing is known except that it is unlikely to contain certain constituents because of its historical background. In other words, there is generally something known about the specimen, if only what it cannot be. This information will be much more important when attempting a full compositional analysis of an unknown specimen, using electron diffraction (§ 6.3) but it is well to bear it in mind even during structure analysis.

6.2.1 STRUCTURE ANALYSIS – SINGLE CRYSTAL SPECIMENS

An important requirement in structure analysis by electron diffraction in the electron microscope is a specimen tilting stage, and preferably one which tilts to a high angle, that is, at least $\pm 30°$, about any axis in the horizontal plane. This is necessary, because, as was seen in § 4.7.2, a single crystal diffraction pattern is only a single section through the reciprocal lattice, and thus is a record of only two dimensions of the three-dimensional crystal lattice. In order to observe the third dimension of the lattice, it would be preferable to tilt the specimen through 90°, about an axis in the plane of the specimen, and again form a diffraction pattern. Such a procedure, to form two diffraction patterns of the crystal exactly 90° apart, would generally give all the information necessary to allow a unique determination of the crystal structure. However, of course, 90° of specimen tilt from the horizontal is not possible in the electron microscope, although $\pm 60°$ tilt stages are now commercially available (Lucas 1970; Philips technical publication, WA6-D10)

The general technique for structure analysis is to tilt the crystalline specimen randomly until a well-defined symmetrical diffraction pattern is formed, record the pattern photographically, and also the specimen tilt angle and axis. Then further tilt the specimen, about the axis of one of the major zones, that is, row of spots, in the diffraction pattern, until a further well-defined diffraction pattern occurs. Again, record the diffraction pattern, and the tilt angle and axis. This procedure should be repeated for higher tilt angles and further major zone tilt axes, to record as many different diffraction patterns, as the limits of the specimen tilting stage permit.

As the specimen height, and thus the objective lens focal length and the camera length, L, will vary during these tilting experiments, the specimen should be coated with a thin layer of a standard, possibly gold or thallous chloride (§ 5.1.3), for an accurate analysis. A polycrystalline ring diffraction pattern will then be superimposed on each of the diffraction patterns, and may be used to give an accurate calibration of the camera constant, λL, for analysis of each diffraction pattern. This repeated calibration procedure may be obviated by the use of an accurate axis-centered tilt stage, for which the specimen height does not vary during tilting, but the camera constant must still be calibrated by coating the specimen, for each series of diffraction patterns taken from one specimen area.

It then remains to analyse the series of diffraction patterns and determine the specimen crystal structure, as described in § 4.8. As this may be a fairly complex problem, if no well-defined diffraction patterns are found which match one of the common reciprocal lattice sections (Appendix 4), any prior knowledge of the specimen may be very useful at this stage. Of the fourteen crystal lattices, a number are fairly uncommon among certain materials. This is particularly true among the metallic elements, which generally crystallize in an f.c.c., b.c.c. or hexagonal structure. Then the use of the procedure described in § 4.8 or the comparison of the diffraction patterns with the common reciprocal lattice sections for these structures, should allow a rapid determination of the crystal structure.

6.2.2 STRUCTURE ANALYSIS – POLYCRYSTALLINE SPECIMENS

By its very nature, a polycrystalline specimen contains single crystal grains at a multitude of random orientations, and thereby contains simultaneously all the information which must be gathered by the successive tilting experiments for a single crystal specimen. In general, structure analysis is easier and more reliable using a polycrystalline specimen, as the complete range of d-spacings is more likely to appear in a polycrystalline diffraction pattern than in the limited selection of diffraction patterns available from a single crystal within the range of the specimen tilting stage.

The procedure is to measure the R value for each diffraction ring, generally by measuring the diameter and dividing by two, and converting these to d-spacings, using the calibrated value of the camera constant, λL. The use of a polycrystalline standard for calibration of the camera constant is now most undesirable, as the two sets of concentric rings, from the specimen and the standard, would be hopelessly intermingled. The standard calibration materials do not condense as a single crystal, unless deposited on a particular substrate, and the technique for forming a single crystal standard, for example gold, is to evaporate the gold on to a heated rock-salt substrate (Pashley 1956). The rock-salt is dissolved away in water and a small piece of the gold standard is placed next to the specimen on the grid, ensuring that the specimen and standard are the same height by using a second grid to create a sandwich. The calibration diffraction pattern can now be taken by moving the specimen stage over the standard specimen area. Sufficient accuracy in the determination of the camera constant, λL, will be possible, provided that the specimen grids are not buckled or sagging, and that the specimen plane is either horizontal, or tilted only about an axis close to the line joining the specimen area and the standard area.

Compilation of the complete list of d-spacings, as required in § 4.8, is now possible, and determination of the specimen crystal structure can proceed as in § 4.8.

6.2.3 EVALUATION OF SPECIMEN DAMAGE

The extent of specimen damage in the electron microscope has recently been highlighted by the application of both electron and optical diffraction techniques. While optical diffraction of the recorded photographic plates may give an accurate measure of the crystallinity of the specimen, subsequent to its examination in the microscope, electron diffraction may be used to monitor any changes in specimen crystallinity that may occur during observation. The immediacy of electron diffraction analysis is very important, although optical diffraction does have the advantage that a very much smaller specimen area may be analysed than the ~1 μm diameter area, possible with electron diffraction.

Specimen damage may occur in the electron microscope by a variety of mechanisms. Specimen heating is a problem well known to electron microscopists, but in addition damage may occur both by the bombardment of the specimen by gaseous ions formed in the microscope and by the incident electron beam itself. Use is made of high vacuum and high-voltage microscopes, image intensifiers and low temperature stages to reduce the rate of specimen damage, but rarely are they successful in eliminating damage of sensitive materials entirely. Metallic specimens only undergo damage at high voltages, inorganic minerals and materials may be damaged at conventional voltages, and organic and biological crystals are definitely known to suffer damage in the electron microscope (Glaeser and Thomas 1969; Glaeser et al. 1970; Thach and Thach 1970; Glaeser 1971).

Electron diffraction is a considerably more sensitive monitor of specimen damage than high-resolution imaging, as the presence of periodic structures down to 0.1 nm or less is easily detected by diffraction analysis. The existence of sharp diffraction spots or rings implies the maintenance of well-defined crystal structure, down to the level of the outermost spots or rings in the diffraction pattern. Determination of the *d*-spacing of the outermost sharp diffraction establishes the limit of preservation of periodic structure in the specimen. As specimen damage occurs, the outermost diffraction spots may be broadened or streaked, or converted into rings, or may simply disappear altogether. Conversely, initially amorphous materials may be crystallized in the electron beam, and diffuse diffraction rings become sharper as 'damage' occurs.

In a review of radiation damage of biological material, Stenn and Bahr (1970) suggest that damage may not affect the meaningful information obtainable, except at spacings of less than 0.4 nm. Glaeser (1971), has studied the radiation damage of three crystalline materials (a) the amino acid l-valine, analogous to proteins, (b) adenosine, analogous to the nucleotide-phosphate and nucleic acid structures, and (c) catalase, as the best known high resolution test specimen that can be prepared by present techniques. The diffraction analysis indicates that the 'best resolution' to be expected, as a result of the limitation of radiation damage is approximately (a) 5 nm for l-valine, (b) 2 nm for adenosine, and (c) 1.5 nm for catalase. However, Glaeser does show, that under carefully controlled conditions, periodicities of 0.82 nm spacing may be observed in a catalase crystal, although this must be considered as probably the highest resolution achieved to-date in a biological specimen.

6.3 Composition analysis

The final analysis for which electron diffraction may be used is the determination of the chemical composition of an unknown crystalline specimen. This is possible, under favourable circumstances, by comparing the *d*-spacings of the unknown specimen with those determined from known standard specimens by previous X-ray or electron diffraction analyses. These *d*-spacings are compiled in tables of crystallographic data, particularly the A.S.T.M. index, Donnay et al. (1963), Wycoff (1963) and the International Tables for X-ray Crystallography.

The procedure for the compositional analysis of an unknown specimen is to compile a list of its *d*-spacings, by measuring all the rings in a polycrystalline diffraction pattern or by forming as many different single crystal patterns as is possible within the limits of the available specimen tilting stage, and determining the *d*-spacings from all the different *R* values obtained.

Compositional analysis is then attempted by searching through the A.S.T.M. index to find a material with a set of *d*-spacings which match those of the unknown specimen. The search may be considerable foreshortened by any knowledge of what the specimen may or may not be, and particularly by any knowledge of the elemental constituents of the specimen gained by X-ray microanalysis, as outlined in § 4.8.

In practice, the composition of the crystalline specimen is rarely entirely unknown, and may at best, be one of only two possible materials, when of course differentiation is fairly simple.

6.3.1 ANALYSIS OF FINE CRYSTALLINE PARTICLES IN BIOLOGICAL MATERIAL

An example of the application of electron diffraction to the analysis of biological material is given by the study of Liabson and Damiano (1969), who differentiated between two possibilities for the composition of the fine crystalline particles present in conjunctival and corneal biopsies and in sternal marrow aspirate taken from a patient with a paraproteinemia. The crystals were analysed by selected area electron diffraction, and determined as chloresteryl stearate rather than cystine, by comparison of the measured *d*-spacings with those tabulated in the A.S.T.M. index.

In addition, electron diffraction may be used to confirm the presence of a suspected material, either by a direct comparison of the diffraction patterns from the 'unknown' specimen and from a prepared standard, or else by determination of the *d*-spacings of the 'unknown' material and comparison with available data. Examples of this type of analysis are (i) the confirmation by Ben-Shaul and Shmueli (1968) that the crystal-like bodies isolated from both carrot roots and tomato fruits are crystals of carotene, by a comparison of the diffraction patterns obtained from the isolated crystals and from a specimen recrystallized from commercial carotene, and

(ii) the confirmation by Tandler and Solari (1969) that the crystalline deposits in the nucleolar areas of maize root cells fixed in lead acetate are lead orthophosphate, by a comparison of the diffraction patterns from the crystalline deposits and a standard sample of lead orthophosphate.

Murray and Ferrier (1968) have described the application of electron diffraction to the study of catalase and ferritin crystals and collagen fibres, and Helander (1971) has used low angle diffraction techniques in the study of crystalloids with large subunits in biological material.

6.3.2 ANALYSIS OF ENZYME HISTOCHEMICAL REACTION PRODUCTS

A further application of electron diffraction techniques is the analysis of the products of enzyme histochemical reactions. For example, Arstila et al, (1966) obtained electron diffraction patterns from cytosomes in glutaraldehyde-fixed rat kidney tissue, and confirmed the composition of the electron dense final reaction product of a demonstration of aryl sulphatase activity by reaction with barium chloride, as barium sulphate. Thus it was confirmed that the heavy staining of the cytosomes could not be due to non-specific binding of barium ions, but indicated a prior liberation of sulphate ions, which could only be effected by the enzymatic action. Though the technique is simple and rapid, it suffers from the disadvantage that diffraction patterns can only be obtained from fairly large organelles having rather abundant reaction products, such as cytosomes.

REFERENCES

Andrews, K. W., D. J. Dyson and S. R. Keown (1971), Interpretation of electron diffraction patterns (Hilger and Watts, London).

Arstila, A. U., S. Jaakkola, H. Kalimo, H. Helminen and V. K. Hopsu-Havu (1966), Electron diffraction as a control of enzyme histochemical reactions at the ultrastructural level, J. Microscop. *5*, 777.

Ashby, M. F. and L. M. Brown (1963), Diffraction contrast in spherically symmetric coherency strains, Phil. Mag. *8*, 1083; On diffraction contrast from inclusions, Phil. Mag. *9*, 1649.

A.S.T.M. Index, Index to the powder diffraction file (Am. Soc. for Testing Mater., Philadelphia, U.S.A.).

Barrett, C. S. (1952), Structure of metals (McGraw-Hill, New York).

Ben-Shaul, Y. and U. Shmueli (1968), Electron diffraction of carotenoid pigments in chromoplasts, Proc. 4th Eur. Reg. Cong. on Electron Microscopy, Rome, p. 411.

Cockayne, D. J. H., J. R. Parsons and C. W. Hoelke (1971), A study of the relationship between lattice fringes and lattice planes in electron microscope images of crystals containing defects, Phil. Mag. *24*, 139.

Donnay, J. H., G. Donnay, E. G. Cox, O. Kennard and M. V. King (1963), Crystal data determinations tables, A.C.A. Monograph No. 5.

France, L. K. and M. H. Loretto (1968), Influence of size of diffracting vector on the visibility of dislocations in α-iron, Proc. Roy. Soc. *A307*, 83.

Gevers, R (1963), On the kinematical theory of diffraction contrast of electron transmission microscope images of perfect and partial screw dislocations, Phil. Mag. *8*, 769.

Glaeser, R. M. and G. Thomas (1969), Application of electron diffraction to biological electron microscopy, Biophys. J. *9*, 1073.

Glaeser, R. M., T. F. Budinger, P. M. Aebersold and G. Thomas (1970), Radiation damage in biological samples, Proc. 7th Int. Cong. on Electron Microscopy, Grenoble *1*, 463.

Glaeser, R. M. (1971), Limitation to the significant information in biological electron microscopy as a result of radiation damage, J. Ultrastruct. Res. *36*, 466.

Groves, G. W. and M. J. Whelan (1962), The determination of the sense of the Burgers' vector of a dislocation from its electron microscope images, Phil. Mag. 7, 1603.

Helander, H. F. (1971), Low angle electron diffraction – a tool for the study of crystalloids with large subunits, J. Ultrastruct. Res. *36*, 527.

Hirsch, P. B., A. Howie and M. J. Whelan (1960), A kinematical theory of diffraction contrast of electron transmission microscope images of dislocations and other defects, Phil. Trans. Roy. Soc. *A252*, 499.

Hirsch, P. B., A. Howie, R. B. Nicholson, D. W. Pashley and M. J. Whelan (1965), Electron microscopy of thin crystals (Butterworths, London).

International tables for X-ray crystallography (Kynoch Press, Birmingham, U.K.), 1952 onwards.

Johari, O. and G. Thomas (1969), Stereographic projection and its application (Interscience Publishers, New York).

Laird, C., E. Eichen and W. R. Bitler (1966), Accuracy in the use of electron diffraction spot patterns for determining crystal orientation, J. Appl. Phys. *37*, 2225.

Liabson, P. R. and V. V. Damiano (1969), X-ray and electron diffraction of ocular and bone marrow crystals in paraproteinemia, Science *163*, 581.

Lucas, J. H. (1970), A 60° double tilt stage for the AEI EM802, Proc. 7th Int. Cong. on Electron Microscopy, Grenoble *1*, 159.

Murray, R. T. and R. P. Ferrier (1968), Biological applications of electron diffraction, J. Ultrastruct. Res. *21*, 361.

Otte, H. M., J. Dash and H. F. Schaake (1964), Electron microscopy and diffraction of thin films: interpretation and correlation of images and diffraction patterns, Phys. Stat. Sol. *5*, 527.

Pashley, D. W. (1956), The study of epitaxy in thin surface films, Advan. Phys. *5*, 173.

Philips technical publications, WA6-D10, Goniometer stage for EM200 and EM300 electron microscopes.

Phillips, F. C. (1955), An introduction to crystallography (Longmans, London).

Ryder, P. L. and W. Pitsch (1968), On the accuracy of orientation determination by selected-area diffraction, Phil. Mag. *18*, 807.

Shenin, S. S. (1965), Method for the precise determination of orientation from electron diffraction patterns, J. Appl. Phys. *36*, 3267.

Shenin, S. S. and C. D. Cahn (1965), The determination of orientation from Kikuchi patterns, Phys. Stat. Sol. *11*, K1.

Silcock, J. and W. J. Tunstall (1964), Partial dislocations associated with NbC precipitation in austenitic stainless steel, Phil. Mag. *10*, 361.

Steeds, J. and P. B. Hirsch (1963), Report on symposium on relation between structure and strength of metals and alloys N.P.L., Teddington, H.M. Stationary Office, London).

Stenn, K. and G. H. Bahr (1970), Specimen damage caused by the beam of the transmission electron microscope; a correlative reconsideration, J. Ultrastruct. Res. *31*, 526.

Tandler, C. J. and A. J. Solari (1969), Nucleolar orthophosphate ions: electron microscope and diffraction studies, J. Cell Biol. *41*, 91.

Thach, R. E. and S. S. Thach (1970), Damage to biological samples caused by the electron beam, Proc. 7th Int. Cong. on Electron Microscopy, Grenoble *1*, 465.

Thomas, G. (1962), Transmission electron microscopy of metals (J. Wiley, New York).

Von Heimendahl, M., W. Bell, and G. Thomas (1964), Application of Kikuchi line analyses in electron microscopy, J. Appl. Phys. *35*, 3614.

Wood, E. A. (1963), Crystal orientation manual (Columbia University, New York).

Wycoff, R. W. G. (1963), Crystal structures 1–6 (J. Wiley, New York).

FURTHER READING

Cowley, J. M. (1967), Crystal structure determination by electron diffraction, Progr. Mater. Sci. *13*, No. 6.

Vainshtein, B. K. (1964), Structure analysis by electron diffraction (Pergamon Press, Oxford, U.K.).

Zvyagin, B. B. (1967), Electron diffraction analysis of clay mineral structures (Plenum Press, New York).

Chapter 7

Information contained in the fine structure of the diffraction pattern

In addition to the discrete spots which occur in the electron diffraction pattern from a single crystal specimen, fine structure may also appear which may be analysed to determine the exact orientation of the specimen, the shape of the crystallites in the specimen or the presence of twins and layered structures. This fine structure may take the form of lines crossing the pattern, streaking out of the diffraction spots or the presence of additional spots or fine satellite reflections around existing spots. Also, the broadening of diffraction spots or of the rings in polycrystalline diffraction patterns may reveal the size of the crystallites in the specimen.

7.1 Kikuchi lines

Diffraction patterns from thick specimens often contain, in addition to the diffraction spots, complex patterns of pairs of dark and bright lines, which are seen in contrast against the diffuse background scattering. These lines are called Kikuchi lines after their discoverer (Kikuchi 1928).

7.1.1 ORIGIN OF KIKUCHI LINES

Kikuchi lines arise from the Bragg diffraction of electrons which were initially inelastically scattered on entering the top surface of the specimen. In addition to the electrons which pass straight through the crystal and those that are diffracted by particular lattice planes without loss of energy (termed elastic scattering), a proportion of the electrons are inelastically scattered through a wide range of angles (§ 3.1), losing energy in the process. Some of these electrons, which are scattered over a wide range of angles, will be travelling at such an angle that they are subsequently elastically scattered by lattice planes lying at a Bragg angle, θ. Fig. 7.1 shows that electrons inelastically scattered in the direction OA will be diffracted into a direction parallel to OB, while electrons scattered in a direction OB will then be diffracted into a direction parallel to OA. If the numbers of electrons initially inelastically scattered in the two directions OA and OB were exactly equal,

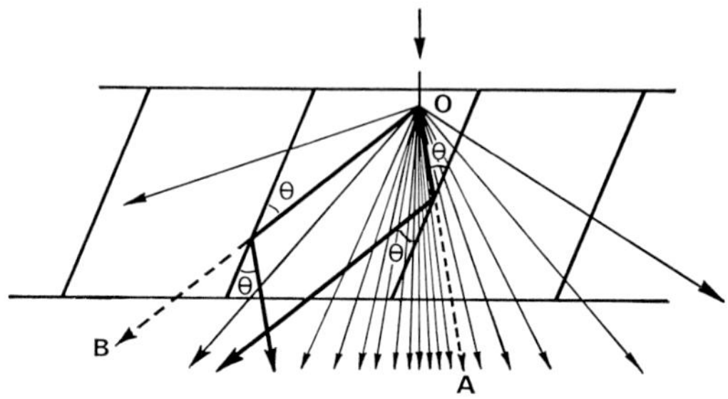

Fig. 7.1. The origin of Kikuchi lines.

then following the Bragg diffraction there would be no net gain or loss of electrons travelling parallel to either of the two directions OA and OB. However, the wider the angle of inelastic scattering the fewer are the electrons scattered (§ 3.1). So following the initial inelastic scattering, there are more electrons travelling in direction OA than in direction OB, but following the Bragg diffraction there is now a net loss of electrons travelling parallel to direction OA and a net gain of electrons travelling parallel to direction OB. The electrons are actually scattered into cones which will intersect the screen or photographic plate as slightly curved lines, the net loss line, or 'defect' line, appearing dark on the viewing screen against the diffuse background, and the net gain line, or 'excess' line, appearing bright on the screen. Thus Kikuchi lines always appear as a pair of dark and bright lines.

Further similar pairs of lines will appear in the diffraction pattern for all the other possible reflecting planes in the crystal, so that the inelastically scattered electrons give rise to a complex array of Kikuchi line pairs, which can be indexed with the Miller indices of the lattice planes from which they originate, just as were the individual diffraction spots.

7.1.2 INDEXING THE KIKUCHI LINES

The angle between OA and OB is 2θ, where θ is the Bragg angle which satisfies the equation

$$\lambda = 2d_{hkl} \sin \theta.$$

So the distance between the dark and bright lines in the Kikuchi line pair,

$$S = 2\theta L = \frac{\lambda L}{d_{hkl}},$$

since for small angles θ, $2\theta \approx 2 \sin \theta$.

If the camera constant, λL, is determined by calibration with a known specimen standard, d-spacings may be determined from the measured values of S, and indices assigned, as in §§ 4.2 and 4.8.

The spacing, S, between the dark and bright lines in a Kikuchi line pair, will be equal to the distance between the cental undeviated spot in the diffraction pattern, and the diffraction spot with the same Miller indices as those assigned to the Kikuchi line pair. So that if one line of the pair lies through the central spot, the other line will cross the associated diffraction spot. A line drawn midway between a Kikuchi line pair represents the intersection of the particular lattice planes with the viewing screen or photographic plate. If this intersection passes through the central spot, that is, the dark and bright line are equidistantly spaced on either side of the central spot, parallel lines of equal intensity actually appear, with a band of enhanced intensity in between. The line spacing may be analysed as before, but the intensity of the lines cannot be explained simply, and reference should be made to Thomson and Cochrane (1939) and Pinsker (1953) for a fuller description of this effect.

As a line drawn midway between a pair of Kikuchi lines represents the intersection of a particular set of lattice planes with the photographic plate, it is possible to determine the exact orientation of the crystal when two pairs of Kikuchi lines are present in the diffraction pattern,

7.1.3 EXACT SPECIMEN ORIENTATION DETERMINATION

Following the indexing of a single crystal diffraction pattern, the approximate orientation of the specimen, with respect to the electron beam, may be eatablished by determining the zone axis $[UVW]$ of the diffraction pattern. However, small changes in the orientation of the specimen may not alter the spatial arrangement but only the intensities of the diffraction spots, and although these intensity changes may sometimes be analysed to give a more accurate determination of specimen orientation, this analysis will generally yield an answer accurate to only about one degree, at best. To allow the accurate determination of specimen orientation, for example, in the analysis of grain misorientation across a low-angle boundary, a more precise method is necessary.

While diffraction spots only change in intensity when a crystal is tilted, the Kikuchi line pairs traverse the diffraction pattern as though they were rigidly fixed to the specimen. The reason for this may be seen in Fig. 7.1, as, in the wide cone of inelastically scattered electrons, there are always scattered electrons available to be diffracted whatever the orientation of the lattice planes. So as the lattice planes are tilted, electrons inelastically scattered at differing angles are Bragg diffracted, and the Kikuchi lines appear to sweep across the diffraction pattern. Geometrically, it can be seen that if a line drawn midway between the Kikuchi line pair represents the intersection of the particular lattices planes with the photographic plate, then as the lattice planes are tilted the intersection must sweep across the photographic plate, as though rigidly fixed to the specimen.

A tilt of the specimen, ϕ, about an axis in a plane parallel to the photographic plate will cause a shift of the whole Kikuchi line pattern in a direction at right angles to the same axis, by a distance, $L\phi$, where L is the camera length. Thus, by measurement of the amount and the direction of the shift of the Kikuchi line pair from the symmetry position, either side of the central spot, an accurate determination of the angle and axis of tilt from the exact zone axis $[UVW]$ is possible. If L, the camera length, is 500 mm, a shift of 9 mm in the Kikuchi line pattern will correspond to $\phi = 1$ degree, so an accuracy of about 0.1 degree may be readily achieved in orientation determinations using Kikuchi line analysis. The procedure and accuracy of the analysis have been discussed in detail by Von Heimendahl et al. (1964), Hirsch et al. (1965), Ryder and Pitsch (1968) and Andrews et at. (1971).

7.1.4 DEVIATION FROM THE BRAGG ANGLE

The use of the electron diffraction pattern in setting up the 'two-beam' condition for accurate defect contrast analysis was described in § 6.1.3. In practice, the images of defects are usually observed in optimum contrast in the light areas either side of the dark band which occurs when the lattice planes are set exactly at the Bragg angle. These light image areas appear because the lattice planes in these areas of the specimen are now oriented at a slight deviation from the Bragg angle. The direction, whether positive or negative, and the extent of this 'deviation from the Bragg angle', s, is required for the accurate contrast analysis of lattice displacements, and may be determined by inspection of the Kikuchi line pattern with respect to the diffraction spot pattern.

For lattice planes oriented exactly at a two-beam Bragg condition, the associated Kikuchi line pair will be positioned with the bright line passing through the diffracted spot and the dark line passing through the central spot of the diffraction pattern (Fig. 7.2a). If the lattice planes are now tilted away from the optic axis, the bright Kikuchi line moves outside the associated diffraction spot (Fig. 7.2b), and the deviation parameter, s, is defined as positive. Conversely, if the lattice planes are tilted towards the optic axis (Fig. 7.2c), the bright line appears inside the associated diffraction spot, and s is

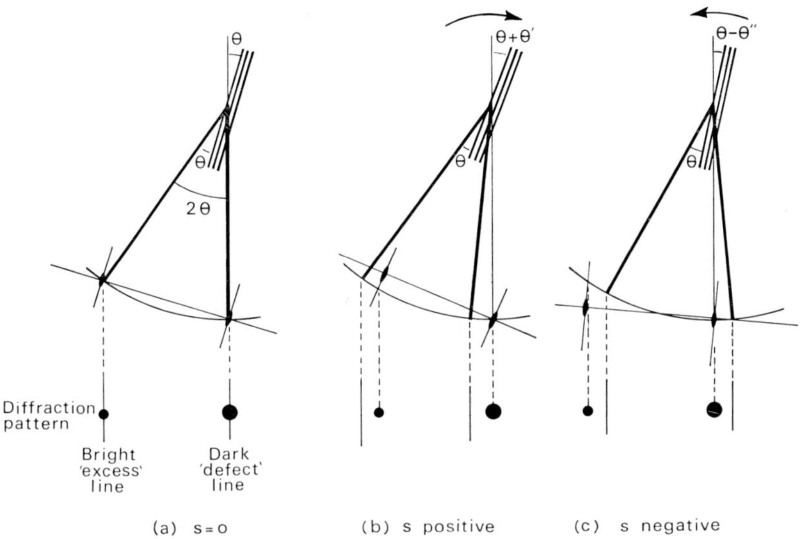

Fig. 7.2. Definition of the sign of s, the 'deviation from the Bragg angle'.

defined as negative, in diffraction contrast experiments on crystalline specimens (Howie and Whelan 1961, 1962; Groves and Whelan 1962). Thomas (1964) has shown that the sign of s can readily be used in contrast analysis experiments to determine the sense of lattice displacements that surround defects.

7.2 Crystal shape effects

If a crystal is thin in one direction the points in the reciprocal lattice representing each set of lattice planes will not be infinitesimally small, but will be extended as reciprocal lattice 'spikes' in the direction of thinness of the specimen (§ 4.7.2). When the specimen is oriented so that the electron beam is parallel to the direction of thinness in the specimen, the Ewald sphere is able to intersect many of the spikes, and cause many more spots to appear in the diffraction pattern than would otherwise be the case if the sphere had to intersect the actual reciprocal lattice points (Fig. 4.21b).

The length of the reciprocal lattice spikes, in a particular direction, is inversely proportional to the thickness of the crystal in that direction.

7.2.1 EXTREMELY THIN CRYSTALS NORMAL TO THE ELECTRON BEAM

If the specimen is extremely thin, the spikes may be sufficiently long to be continuous between the many layers of the reciprocal lattice, and then extra diffraction spots may appear in the diffraction pattern, at intermediate positions, for particular crystal structures in certain orientations. For example, the {110}, {310}, {510}, ... reflections, which are normally forbidden in the [001] diffraction pattern of a face-centred cubic crystal, may appear if the specimen is extremely thin.

7.2.2 PLATELET CRYSTALLITES PARALLEL TO THE ELECTRON BEAM

If the electron beam is perpendicular to the direction of thinness of the crystal, that is, the specimen contains small platelet crystallites lying parallel to the electron beam, the reciprocal lattice points are spiked in the plane of the reciprocal lattice section. The Ewald sphere will now intersect not only the reciprocal lattice points but also the length of the spikes, and the diffraction spots will be streaked parallel to the direction of thinness of the crystallites. A typical example of this effect arises from the precipitation of Guinier-Preston zones in aluminium-4% copper alloys. These zones consist of platelets of copper, one or two atoms thick, lying in {100} planes. When the specimen is oriented with the electron beam parallel to these planes, the spots in the diffraction pattern are streaked in the [100] and [010] directions (see Thomas 1962, Fig. 24).

7.2.3 NEEDLE-SHAPED CRYSTALS

A needle-like crystal has its reciprocal lattice points extended as small discs, the plane of each disc being perpendicular to the long axis of the needle. If the specimen is oriented with the long axis of the crystal perpendicular to the electron beam, the diffraction spots will be streaked perpendicular to the long axis of the needle. When the needle axis is parallel to the electron beam the Ewald sphere intersects the plane of the reciprocal lattice discs and the diffraction spots will be extended as discs of intensity. However, because of the thinness of the reciprocal lattice discs, very few intersections of the

Ewald sphere with the discs actually occur, and this type of diffraction phenomenon is only rarely observed.

7.2.4 SMALL POLYHEDRAL CRYSTALS

Small polyhedral crystals have reciprocal lattices in which the lattice points are spiked in directions perpendicular to each face of the crystal. Thus a small octohedral crystal specimen has reciprocal lattice points with eight symmetrically arranged spikes. If the Ewald sphere intersects the spikes at a shallow angle the diffraction spots will have 'tails', but if the intersection occurs at a larger angle 'splitting' of the diffraction spots may occur. This phenomenon has been observed by Cochrane (1936) in reflection diffraction patterns from cobalt crystals formed on the $\{110\}$ faces of a copper single crystal.

The fine structure of the diffraction pattern produced by the various crystal shapes is discussed in further detail by Hirsch et al. (1965) and Thomas et al. (1965).

7.3 Crystal size effects–spot/line broadening

Just as the geometry of the reciprocal lattice spikes can be analysed to determine the shape of crystallites in the specimen (§ 7.2), so the extent of reciprocal lattice spikes, revealed as spot elongation in the single crystal pattern, or line broadening in the polycrystalline diffraction pattern, may be analysed to determine the size of the crystallites. The extent of the reciprocal lattice spikes in any one direction is inversely proportional to the thickness of the crystal in that direction (§ 4.7.2). Consequently, the broadening of diffraction spots or rings is a measure of the mean dimension of the crystals perpendicular to the diffracting planes (hkl), provided that no lattice strain effects (Blackman 1951; Hirsch et al. 1965), or refraction effects (§ 7.6), are present.

A relationship between crystallite size and X-ray line broadening was first derived by Scherrer (1918), as

$$D = \frac{K\lambda}{\beta \cos \theta_{hkl}},$$

where D is the mean dimension of the crystallites perpendicular to the diffracting planes, that is, in the plane of the thin specimen, K is a constant, λ the electron wavelength, β is the half-width of the broadening due to the specimen and θ_{hkl} the Bragg angle for the particular diffraction spot or line (hkl) being measured. The constant K depends largely upon the crystallite shape, the (hkl) indices and the definition of β and D. However, if β and D are defined as above, a value of K of about 0.9 can be taken (Bragg 1949).

Determination of the broadening β due only to the specimen constitutes the major problem associated with crystallite size analysis. Instrumental factors, such as distortion and chromatic aberration, may give spurious broadening, but these can be eliminated by using a high resolution diffraction stage in the electron microscope (§ 5.3), rather than placing the specimen in the standard specimen position. Lattice strain effects may be minimized by ensuring that the specimen is in a well-annealed state, and refraction effects may be accounted for by considering a number of different reflections in the pattern since the magnitude of the broadening or splitting due to refraction varies appreciably for different reflections.

Assuming a neglible contribution of broadening due to instrumental effects, lattice strain and refraction, the half-width of the broadening, β, due to the specimen is given by

$$\beta = \sqrt{B - b^2},$$

where B is the experimentally measured half-width of the spot or line, and b is the half-width of the centre spot in the absence of the specimen. Rymer and Fayers (1958) found that lattice strains are usually present in a specimen, but suggest that the profile of the innermost ring of a polycrystalline diffraction pattern will generally give a fairly reliable indication of the nature of the broadening other than strain broadening. Thus, by determining the broadening of the innermost ring of a polycrystalline pattern, or of the inner spots of a single crystal pattern, a measure of the dimensions of the crystallites in the plane of a thin specimen may be found.

7.4 Dynamical effects–double diffraction

Electrons diffracted from one set of lattice planes may then act as a primary beam of electrons for diffraction from a second set of planes in the specimen. This second set of planes may be the identical set of planes which caused the first diffraction, as electrons diffracted from one set of planes in a crystal are then travelling at the correct angle to be rediffracted back into the undeviated beam. Alternatively, they may be a different set of planes, in the same crystal or in a second phase or in an epitaxial layer deposited on the first crystal.

Rediffraction from the identical set of planes as the initial diffraction, an effect taken into account in the dynamical theory of electron diffraction, can modify the intensity of spots in the diffraction pattern and sometimes gives rise to 'extra' spots, unexpected from consideration of the specimen structure. If reflections $(h_1k_1l_1)$ and $(h_2k_2l_2)$ occur in the diffraction pattern, reflections $(h_1 + h_2, k_1 + k_2, l_1 + l_2)$ and $(h_1 - h_2, k_1 - k_2, l_1 - l_2)$ may also appear by double diffraction. No 'extra' reflections will occur in an f.c.c. diffraction pattern, as $(h_1k_1l_1)$ and $(h_2k_2l_2)$ must be all even or all odd for the f.c.c. structure, and as $(h_1 \pm h_2, k_1 \pm k_2, l_1 \pm l_2)$ reflections also satisfy this rule, they will already be present in the diffraction pattern. In a b.c.c. diffraction pattern $(h_1 + k_1 + l_1)$ and $(h_2 + k_2 + l_2)$ must be even, and so again the vectorial sums and differences satisfy this rule and no extra reflections will appear. However, hexagonal and other structures do not meet these requirements and may show extra reflections in the diffraction pattern due to double diffraction.

Superlattices, or ordered structures in alloys, also give rise to extra reflections in the diffraction pattern. Glossop and Pashley (1959) have shown that extra spots, in the form of satellite spots around the main reflections, appearing in the diffraction patterns of an ordered CuAu II alloy, can be caused by double diffraction. For further treatment of the origin of extra reflections from superlattices, see Hirsch et al. (1965), and Andrews et al. (1971).

When two single crystal layers with slightly different lattice constants are grown superimposed in parallel orientation, that is epitaxially, extra spots may occur in the diffraction pattern when diffracted beams from one lattice act as primary beams for diffraction by the second lattice. The extra spots often occur as satellite reflections positioned around the main spots in a similar array to that of the main spots of the diffraction pattern (Bassett et al. 1958).

7.5 Twin reflections

The phenomenon of twinning is the formation within a crystal of a region with a new and special orientation with respect to the original matrix. This twin orientation is obtained by rotating the matrix through 180°, about a particular orientation, known as the twin axis, such that the twin lattice is the mirror image of the matrix lattice. It is clear that extra reflections will occur in the diffraction pattern of a specimen area containing both matrix and twins. If the matrix and only a single twin are present in the diffracting area, the diffraction pattern may be analysed as the superposition of two diffraction patterns, the one being the mirror image of the second, about a certain direction in the crystal. By finding this direction in the diffraction pattern, the twin axis and the twin plane, perpendicular to the twin axis, may be determined.

When many twins are present in the diffracting area, on different twin planes –for example, the four {111} planes in the f.c.c. structure, are twin planes– extra spots may appear in the diffraction pattern as satellite spots around the main reflections, or else appear to cause the splitting of the matrix diffraction spots.

If the twin platelets are sufficiently thin, or if there is any elastic strain at the twin interface, the reciprocal lattice points will be spiked, giving rise to either streaks or subsidiary reflections. The direction of these streaks is a direct indication of the twin plane itself. Further treatment of the analysis of twin reflections is given by Hirsch et al. (1965), and Andrews et al. (1971).

7.6 Refraction effects

Just as the bending of light rays occurs on entering media with refractive indices not equal to unity, so a crystal has a refractive index for electrons which is slightly greater than unity. This does not lead to any appreciable deviation of the electron beam, unless the electron beam enters the crystal at a very shallow angle. Thus, for a continuous thin film the deviating effect of refraction as the beam enters the specimen is negligible, but for a specimen of small crystalline particles. for example, cube-shaped magnesium oxide crystals on a support film, a significant deviation of the electron beam may occur as it enters or leaves the crystallites through faces almost parallel to the electron beam. Because it is possible for different electrons to pass through different faces of each crystallite, various deviations are possible, and refraction may cause the splitting of diffraction spots into groups of closely spaced satellite spots.

The occurence of refraction effects from small crystallites within a crystalline matrix is rare, because for most combinations of precipitate and matrix the difference in refractive index is small. Thus for most thin film work, of continuous specimens or matrices containing precipitates, the effect of refraction can be ignored. However, refraction effects may be observed in transmission diffraction patterns from isolated crystallite specimens, and appreciable refraction effects may occur in reflection diffraction patterns if the specimen has a smooth surface. The diffracted beams will be refracted towards the 'cut-off' edge of the reflection diffraction pattern (see Fig. 5.9), giving independent satellite spots if the specimen surface is perfectly smooth, and streaks to the diffraction spots if the surface is undulated. The amount of shift of the diffraction spots due to refraction decreases as the distance between a diffracted beam and the 'cut-off' edge increases, and so the semi-circular arcs of a reflection diffraction ring from a very smooth polycrystalline surface become slightly U-shaped. Further explanation of the fine structure of electron diffraction patterns due to the effects of refraction is given by Hirsch et al. (1965).

REFERENCES

Andrews, K. W., D. J. Dyson and S. R. Keown (1971), Interpretation of electron diffraction patterns (Hilger and Watts, London).

Bassett, G. A., J. W. Menter and D. W. Pashley (1958), Moiré patterns on electron micrographs, and their application to the study of dislocations in metals, Proc. Roy. Soc. *A246*, 345.

Blackman, M. (1951), Diffraction from a bent crystal, Proc. Phys. Soc., London, *B64*, 625.

Bragg, L. (1949), The crystalline state, Vol. 1 (G. Bell and Sons, Ltd., London) p. 189.

Cochrane, W. (1936), The structure of some metallic deposits on a copper single crystals as determined by electron diffraction, Proc. Phys. Soc., *48*, 723.

Glossop, A. B. and D. W. Pashley (1959), The direct observation of anti-phase domain boundaries in ordered copper gold (Cu–Au) alloy, Proc. Roy. Soc. *A250*, 132.

Groves, G. W. and M. J. Whelan (1962), The determination of the sense of the Burgers' vector of a dislocation from its electron microscope images, Phil. Mag. *7*, 1603.

Hirsch, P. B., A. Howie, R. B. Nicholson, D. W. Pashley and M. J. Whelan (1965), Electron microscopy of thin crystals (Butterworths, London).

Howie, A. and M. J. Whelan (1961), Diffraction contrast of electron microscope images. II, the developement of the dynamical theory, Proc. Roy. Soc. *A263*, 217.

Howie, A. and M. J. Whelan (1962), Diffraction contrast of electron microscope images, III, results and experimental confirmation of the dynamical theory of dislocation image contrast, Proc. Roy. Soc. *A267*, 206.

Kikuchi, S. (1928), Diffraction of cathode rays by mica, Japan. J. Phys. *5*, 83.

Pinsker, Z. G. (1953), Electron diffraction (Butterworths, London).

Ryder, P. L. and W. Pitsch (1968), On the accuracy of orientation determination by selected-area diffraction, Phil. Mag. *18*, 807.

Rymer, T. B. and F. J. Fayers (1958), The intensity profiles of electron diffraction lines, Phil. Mag. *3*, 1137.

Scherrer, P. (1918), Evaluation of size and internal structure of colloidal particles by means of Roentgen rays. Gottinger Nachr. *2*, 98.

Thomas, G. (1962), Transmission electron microscopy of metals (J. Wiley, New York).

Thomas, G. (1964), Thin films (A.S.M., Cleveland, Ohio), p. 227.

Thomas, G., W. Bell and H. M. Otte (1965), Interpretation of electron diffraction patterns from thin platelets, Phys. Stat. Sol. *12*, 353.

Thomson, G. P. and W. Cochrane (1953), Theory and practice of electron diffraction (Macmillan, London).

Von Heimendahl, M., W. Bell and G. Thomas (1964), Applications of Kikuchi line analysis in electron microscopy, J. Appl. Phys. *35*, 3614.

FURTHER READING

Amelinckx, A., R. Gevers, G. Remaut and Van Landuyt (eds.), Modern diffraction and imaging techniques in materials science, Proc. of the Intern. Summer Course on Mater. Sc., Antwerp, Belgium, 1969.
(North Holland Publ. Comp., Amsterdam, 1970).

Appendix 1

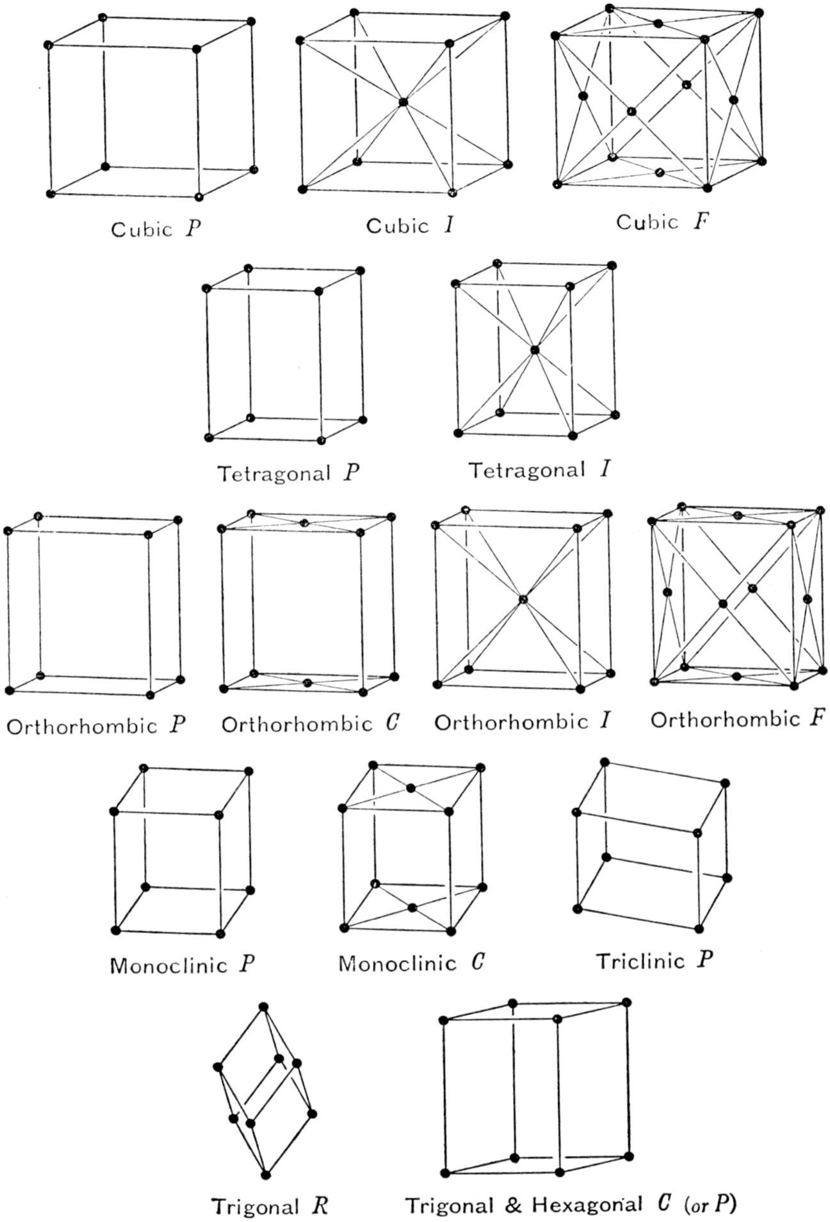

The fourteen crystal lattices

Appendix 2a

The d-spacing formulae for the seven crystal systems

Crystal system	*Interplanar spacing of the* (hkl) *plane*
Cubic	$\dfrac{1}{d^2}=\dfrac{(h^2+k^2+l^2)}{a^2}$
Tetragonal	$\dfrac{1}{d^2}=\dfrac{(h^2+k^2)}{a^2}+\dfrac{l^2}{c^2}$
Orthorhombic	$\dfrac{1}{d^2}=\dfrac{h^2}{a^2}+\dfrac{k^2}{b^2}+\dfrac{l^2}{c^2}$
Hexagonal	$\dfrac{1}{d^2}=\dfrac{4}{3}\left(\dfrac{h^2+hk+k^2}{a^2}\right)+\dfrac{l^2}{c^2}$
Rhombohedral	$\dfrac{1}{d^2}=\dfrac{1}{a^2}\left\{\dfrac{(1+\cos\alpha)[(h^2+k^2+l^2)-(1-\tan^2\alpha/2)(hk+kl+lh)]}{(1+\cos\alpha-2\cos^2\alpha)}\right\}$
Monoclinic	$\dfrac{1}{d^2}=\dfrac{1}{a^2}\left(\dfrac{h^2}{\sin^2\beta}\right)+\dfrac{k^2}{b^2}+\dfrac{1}{c^2}\left(\dfrac{l^2}{\sin^2\beta}\right)-\dfrac{2hl\cos\beta}{ac\sin^2\beta}$
Triclinic	(Complex–see Andrews, K. W., D. J. Dyson and S. R. Keown, 1971, Interpretation of electron diffraction patterns (Hilger & Watts, London).

Appendix 2b

The interplanar angle formulae

Crystal system	*Angle ϕ between $(h_1k_1l_1)$ and $(h_2k_2l_2)$*
Cubic	$\cos\phi = \dfrac{h_1h_2 + k_1k_2 + l_1l_2}{\sqrt{[(h_1^2 + k_1^2 + l_1^2)(h_2^2 + k_2^2 + l_2^2)]}}$
Tetragonal	$\cos\phi = \dfrac{\left(\dfrac{h_1h_2 + k_1k_2}{a^2}\right) + \left(\dfrac{l_1l_2}{c^2}\right)}{\sqrt{\left\{\left[\dfrac{(h_1^2 + k_1^2)}{a^2} + \dfrac{l_1^2}{c^2}\right]\left[\dfrac{(h_2^2 + k_2^2)}{a^2} + \dfrac{l_2^2}{c^2}\right]\right\}}}$
Orthorhombic	$\cos\phi = \dfrac{\dfrac{h_1h_2}{a^2} + \dfrac{k_1k_2}{b^2} + \dfrac{l_1l_2}{c^2}}{\sqrt{\left[\left(\dfrac{h_1^2}{a^2} + \dfrac{k_1^2}{b^2} + \dfrac{l_1^2}{c^2}\right)\left(\dfrac{h_2^2}{a^2} + \dfrac{k_2^2}{b^2} + \dfrac{l_2^2}{c^2}\right)\right]}}$
Hexagonal	$\cos\phi = \dfrac{h_1h_2 + k_1k_2 + \left(\dfrac{h_1k_2 + k_1h_2}{2}\right) + \dfrac{3a^2}{4c^2}l_1l_2}{\sqrt{\left[\left(h_1^2 + k_1^2 + h_1k_1 + \dfrac{3a^2}{4c^2}l_1^2\right)\left(h_2^2 + k_2^2 + h_2k_2 + \dfrac{3a^2}{4c^2}l_2^2\right)\right]}}$
Rhombohedral Monoclinic Triclinic	(Complex–see Andrews, K. W., D. J. Dyson and S. R. Keown (1971), Interpretation of electron diffraction patterns (Hilger and Watts, London).

Appendix 3

Electron wavelength data for a range of accelerating voltages

The relativistic wavelength, λ, is tabulated as a function of the accelerating voltage, V, applied to the electron gun.

Accelerating Voltage V kilovolts, kV	*Relativistic Wavelength λ, nm*
20	0.00859
30	0.00698
40	0.00602
50	0.00536
60	0.00487
70	0.00448
80	0.00418
90	0.00392
100	0.00370
200	0.00251
300	0.00197
400	0.00164
500	0.00142
600	0.00126
700	0.00113
800	0.00103
900	0.00094
1000	0.00087
2000	0.00050
4000	0.00028

Appendix 4

4a Common reciprocal lattice sections

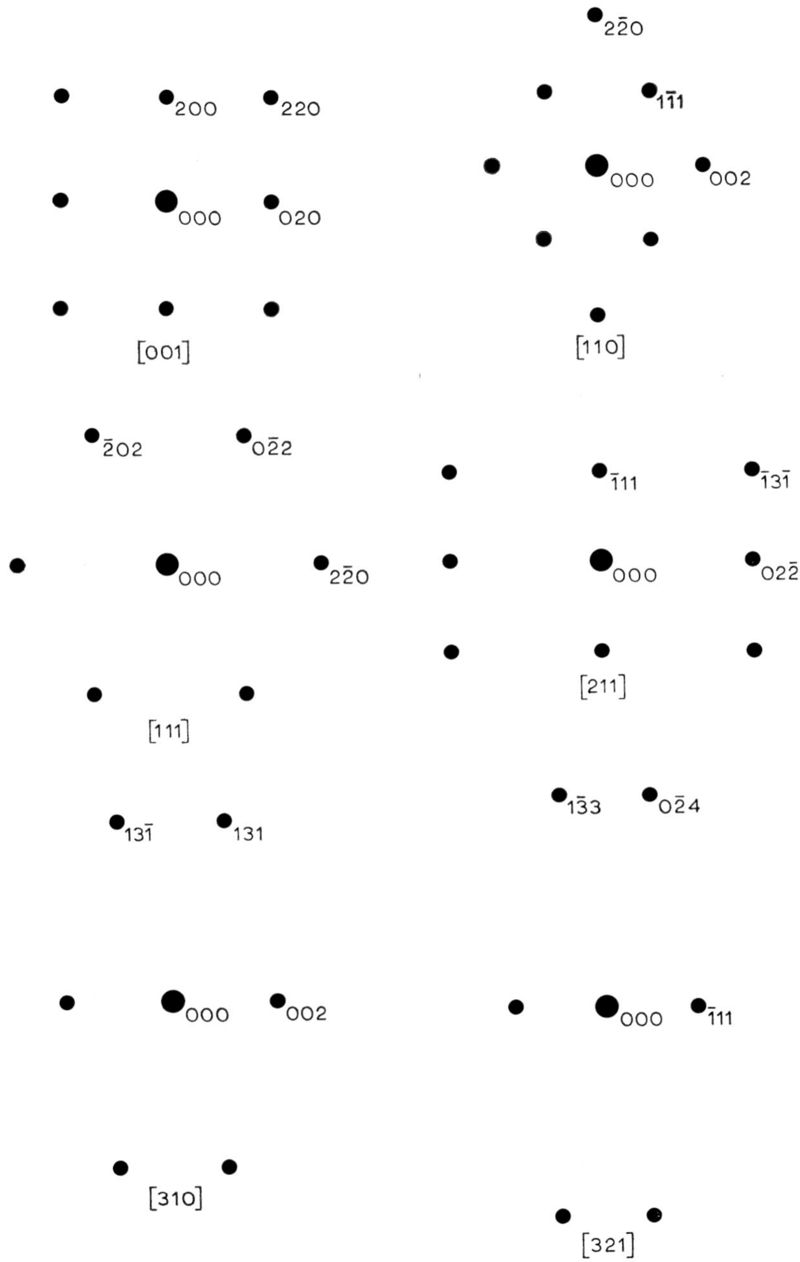

Face-centred cubic structure.

4b Common reciprocal lattice sections

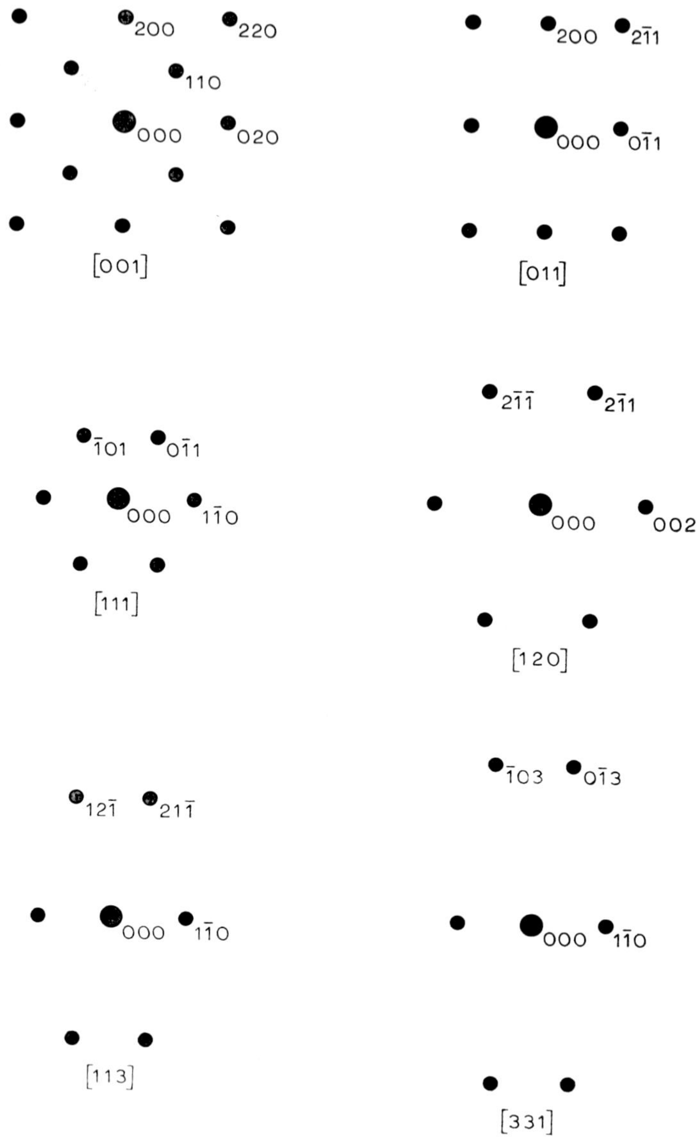

Body-centred cubic structure.

4c Common reciprocal lattice sections

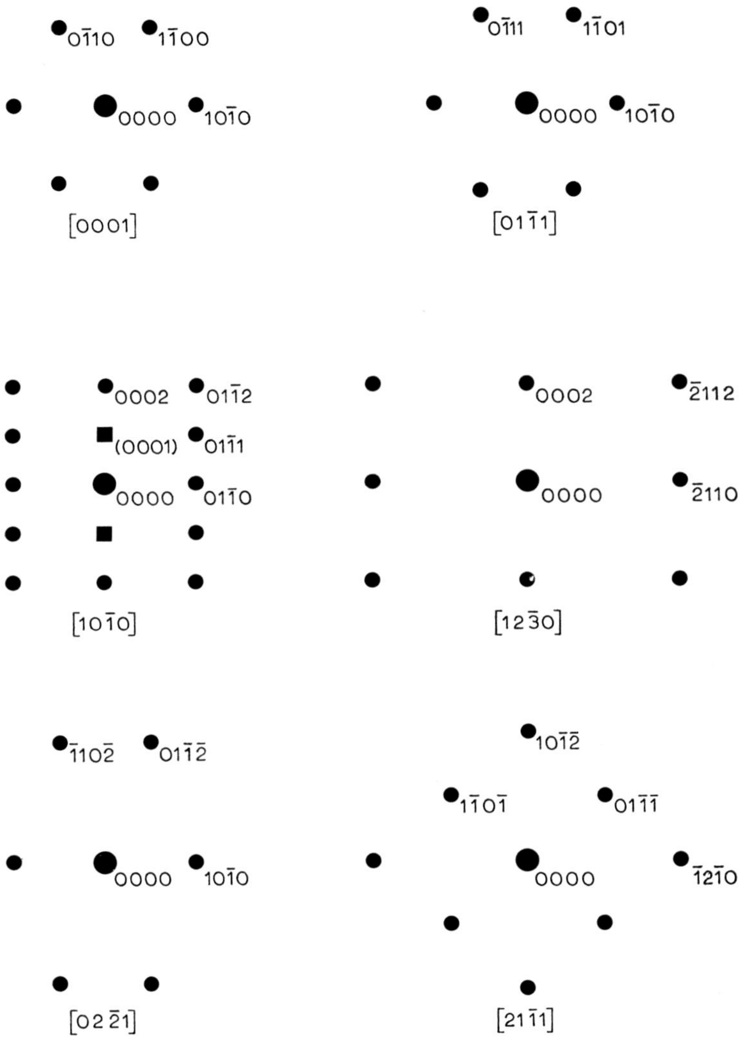

Hexagonal close-packed structure.

2 APPLICATION OF OPTICAL DIFFRACTION AND IMAGE RECONSTRUCTION TECHNIQUES TO ELECTRON MICROGRAPHS

ROBERT W. HORNE and ROY MARKHAM

John Innes Institute

Colney Lane, Norwich

Chapter 1

Introduction

The development of very high resolution electron microscopes of the transmission type has made it possible to obtain a point to point resolution of about 0.3 nm and in the case of certain specimens with regular lattice structures a linear resolution of better than 0.2 nm (see Dowell 1963; Ruska 1965; Yada and Hibi 1968; Thon 1968).

Despite the fact that the performance of the electron microscope is rapidly approaching the theoretical limit, serious problems are encountered in the interpretation of the final image.

1.1 Factors limiting the resolution of detail in electron micrographs

There are limitations in resolving detail (from biological specimens, and in certain non-biological preparations) in electron micrographs due to several basic difficulties which are inherent in the instruments and optical systems available or in current techniques for the preparation of specimens. Specimens which, together with the supporting film, are too thick impose certain limitations on the attainable resolution by producing chromatic changes in the scattered electron beam and electron charge effects in the region of the specimen (Cosslett 1956). For example, the detail visible in a specimen with a total thickness of over 200 nm viewed at a conventional accelerating voltage of 80 kV will be severely limited. The support film also becomes of particular importance when investigating detail in very small particles of low molecular weight where the dimensions of the specimens may be equal to, or smaller than, the thickness of the supporting substrate. Under these conditions the random granulation or 'noise' from the amorphous support film, which is superimposed on the image of the specimen, limits the resolution that is obtainable. It has been well established that the contrast of this random granulation depends on the phase-contrast effects that result when the objective lens is set slightly above or below the true focal position. Moreover, the appearance of fine detail in the final image of the specimen itself will vary according to the phase-contrast conditions and this is of some importance in the analysis of electron micrographs by optical diffraction methods. For a more detailed theoretical discussion of phase-contrast effects and their influence on the electron microscope image the reader is referred to the publications of Haine and Cosslett (1961), Thon (1967, 1968), Heidenreich (1967), Hanzen (1971) and Unwin (1971). In addition to phase-contrast effects there are randomly scattered electrons from the various internal parts of the microscope which are capable of reducing the signal-to-noise ratio. The signal-to-noise ratio in the case of the electron microscope is essentially the same as that encountered in any communications system (Gabor 1965). Noise is generated by the erratic, intermittent, or statistically random arrival of electrons at the final electron image screen or photographic plate. Additional noise can be produced after the image has been recorded by the presence of grain in a given photographic plate or film emulsion.

One of the most serious problems associated with electron microscopy of biological specimens and polymers is that of radiation damage by the illuminating beam in the microscope (Ruska 1965; Reimer 1965; Glaeser and Thomas 1969; Williams and Fisher 1970a, b). When other adverse effects, such as contamination, are added to the problems of resolving small detail from the specimen, it is not surprising to find that much of the basic information recorded on the micrograph is masked or obscured by unwanted noise, or that it has been changed by irradiation effects. Only limited information is available concerning radiation damage since the problem is still relatively new and has received comparatively little attention. It is beyond the scope of our discussion here. The removal of noise, on the other hand, by image filtering techniques is a distinct possibility when the electron microscope is considered in terms of a communication channel as mentioned earlier.

There are several experimental approaches to the problem of image filtering and these can be considered to fall under two headings; the extraction of information before the final electron microscope image is recorded, and filtration of features from the photographic negative. The first of these approaches involves the investigation of electron lens systems which act as velocity filters and can thus select electrons scattered from the specimen with specific energy losses (see Castaing and Lucien 1964). These experiments fall within the field of advanced electron optics and are not within the scope of this monograph. The second practical approach for image filtering is the application of optical diffraction and photographic techniques which make it possible to extract information concerning repetitive features appearing in the final photograph and to analyse their spacings and dimensions.

In this monograph the basic principles of the optical diffractometer will be discussed and how it can be applied to the analysis of electron micrographs, together with its limitations and defects. Other chapters will be concerned with techniques associated with various photographic systems for the averaging of structural features recorded on the plate or negative. Emphasis will be placed on the more practical aspects of optical diffraction, image reconstruction and photographic integration, together with details for constructing suitable apparatus.

1.2 The development of optical diffractometers

It is interesting to note that diffraction techniques have already been applied to physical and biological problems over a number of years and that there are close similarities between the basic principles of X-ray diffraction, electron diffraction and optical diffraction methods for determining fine structure. The basic theory and techniques of electron diffraction are discussed in detail by Beeston in the first part of this monograph. When X-ray diffraction techniques are applied to non-biological or biological crystalline material diffraction patterns are obtained, but due to the lack of any suitable lens systems for X-rays it is not possible to recombine the image subsequently. In contrast, suitable optical systems can be arranged which will provide an analogue as an aid to the interpretation of X-ray patterns (Buerger 1941; Bragg 1944; Wyckoff et al. 1957).

One of the first experiments carried out with the aid of an optical diffractometer was done as early as 1950 by Wilkins et al. in a study of crystalline inclusions of tobacco mosaic virus which were sufficiently perfect to act as optical gratings capable of separating the lines of a mercury arc spectrum. The development of the optical diffractometer for the study of macroscopic patterns designed to represent the projections of atoms arranged in crystal lattice planes was largely due to the work of Lipson and Taylor (Taylor and Lipson 1951; Lipson and Taylor 1951; Taylor and Lipson 1964).

The application of the optical diffractometer to electron microscopy is a relatively new development, although it was pointed out by McLachlan (1958) that techniques derived from X-ray crystallography might be used to enhance certain repeating features in electron micrographs. New interest in the optical diffractometer for the analysis of electron micrographs from biological objects was stimulated by rapid developments in the performance of the electron microscope, coupled with the introduction of negative staining as an improved method of specimen preparation. These developments also coincided with the production of the laser as a high intensity and coherent light source suitable for the diffractometer.

The extraction of information of fine detail from images of small biological particles with the aid of photographic averaging techniques was investigated by Markham et al. (1963, 1964), and it soon became clear that, in order to determine the spacings and angles of repetitive features present in electron micrographs with some degree of accuracy, some method of measurement, such as optical diffraction, was desirable which would give a more direct indication of the rotational or linear spacings recorded on the plate. Independent applications of optical diffraction and filtering techniques to electron micrographs were made by Klug and De Rosier (1966), Bancroft et al. (1967), Markham (1968), and Fraser and Millward (1970). These experiments have led to a rapid expansion of the use of optical diffraction as an aid to the electron microscopist. It should be mentioned here however, that although the optical diffractometer and various photographic averaging techniques offer some scope in image enhancement, the patterns observed are frequently of a complex nature and they require some knowledge of the principles relating to the theory of diffraction in order to analyse and interpret the diffraction patterns (see Taylor and Lipson 1964; De Rosier 1971; Moody 1971).

1.3 What is an optical diffraction pattern?

The formation of an optical diffraction pattern, when a coherent beam of light is passed through a transparent object, is almost exactly analogous to the formation of an electron diffraction pattern by a beam of electrons or an X-ray diffraction pattern by an X-ray beam. The principles of diffraction are identical for all types of radiation, although the practicalities of the mode of formation of the diffraction pattern and of the instrumentation may differ considerably. The electron diffraction camera consists of a source of electrons (the electron gun and electromagnetic condenser lenses), the specimen and possibly (as in the electron microscope) an objective lens and projector lenses for observing the electron diffraction pattern on a phosphor viewing screen. Similarly, the optical diffraction camera consists of a source of radiation, an object and possibly a lens system to observe the diffraction pattern, but in this case the source is usually a helium-neon laser with one or more glass lenses to illuminate the object, and a glass lens system for viewing the optical diffraction pattern which is usually formed on a ground glass screen. The object may be anything that is transparent to light, but in the application of the optical diffractometer to electron microscopy the object is nearly always the photographic plate or film recorded in the electron microscope, or a copy or enlargement of the micrograph on film. The exceptions may be test objects used in the calibration of the camera length of the diffractometer, when in addition to micrographs of periodic objects of known spacing, optical diffraction gratings or metal meshes of known dimensions may be used. Except in these cases, the object will always be referred to in the following text as the micrograph, or photographic plate or film.

Just as in electron diffraction (Beeston, Chapter 1), the form of the optical diffraction pattern is directly related to the structural detail of the micrograph. If the micrograph is of an amorphous specimen area and shows no periodic or regular detail, then the optical diffraction pattern will contain only a diffuse 'noise' spectrum radially distributed about the central diffraction spot. The 'noise' spectrum may be altered by the presence of astigmatism, specimen drift or out-of-focus effects in the microscope image (§ 5.4), and in this way provides information about the recorded micrograph, but it does not provide any information about the structure of the specimen except, of course, that it is amorphous. The extent of the noise spectrum does, however, indicate the limit of resolution of detail that is recorded on the electron micrograph (Fig. 1.1). If the specimen area is crystalline or the micrograph contains periodic detail then, in addition to the noise spectrum, discrete diffraction spots (Fig. 5.3a) or lines (Fig. 1.1) will be observed in the optical diffraction pattern, at distances and orientations from the central diffraction spot that can be analysed to provide information about the periodic structure of the corresponding specimen area.

Specimen areas recorded for optical diffraction analysis are usually single objects with periodic features and the optical diffraction pattern then consists of discrete diffraction spots. Polycrystalline image areas, such as that in Fig. 1.2, rarely occur because of the relatively high magnifications at which images are recorded, and so the occurrence of optical diffraction patterns containing rings is sufficiently rare that it is convenient to refer only to diffraction spots when discussing the diffraction maxima in optical diffraction patterns.

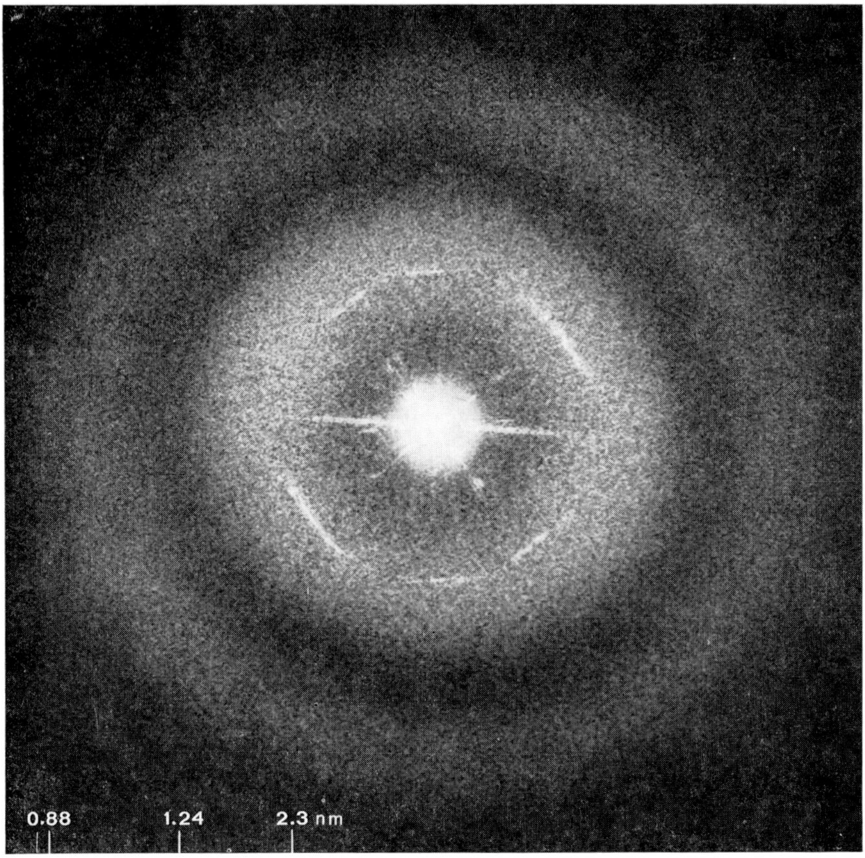

Fig. 1.1. Optical diffraction pattern from an area of an electron micrograph containing the images of several tobacco mosaic virus particles (Fig. 1.2). The prominent white lines arise from the pitch of the virus helix (2.3 nm). Note that the noise spectrum (the diffraction rings) has zero intensity at a radius of 1.24 nm, but extends out as far as 0.88 nm. In this particular micrograph point-to-point measurements might indicate a resolution of 0.88 nm, but since the phase has changed at 1.24 nm, detail outside this region would be spurious. It would be possible to correct for this by a phase-changing plate but detail in the region of the minimum (1.24 nm) is lost irrevocably.

As in electron diffraction, the optical diffraction spots become sharper the more regular and extensive is the periodic array recorded in the micrograph. The optical diffraction pattern can be considered as a coded message giving in position, amplitude and phase all the information about the micrograph as probed by the coherent light source. When this information is recorded, either on the retina, or on a photographic emulsion, or on a photocell, some

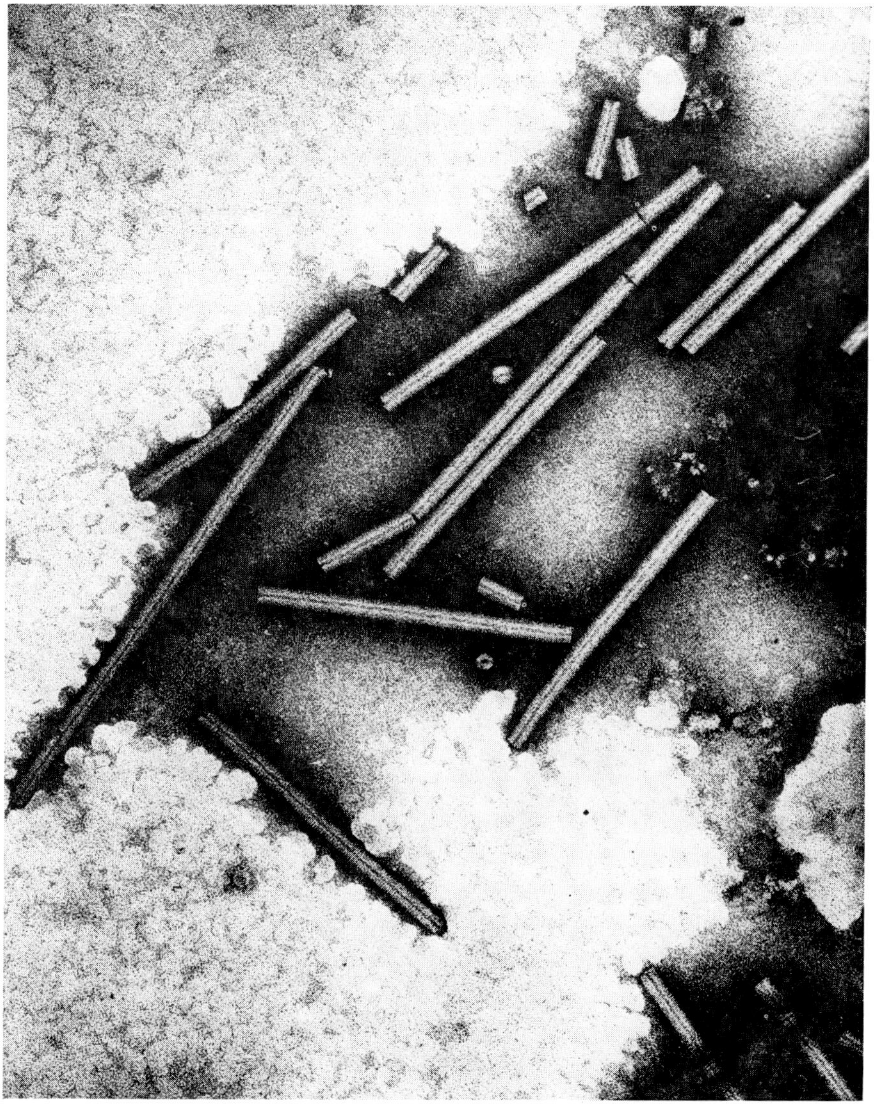

Fig. 1.2. Electron micrograph of a negatively-stained preparation of tobacco mosaic virus rods used to produce the optical diffraction pattern shown in Fig. 1.1. (Reproduced by kind permission of R. C. Williams and H. W. Fisher, see J. Mol. Biol. *52*, 121, 1970.)

of this information is lost. This is because detection systems of the type mentioned do not respond to the phase of light waves, and also only respond to the intensity, the square of the amplitude, rather than the amplitude. Consequently these systems respond equally to positive and negative amplitudes and do not discriminate between them. The net effect is that the optical diffraction pattern as seen or recorded does not locate the object in space, or even whether it is moving with respect to the optical axis of the system, other than rotating in a plane at right angles to the axis. It is also insensitive to whether the object is rotated through 180° in this plane and (within the limits of the necessity for having a limit to its periphery) to whether the object is positive or negative, i.e. dark spots on a light background or light spots on a dark background.

It must always be remembered in optical diffraction analysis that the diffracting object is the electron micrograph and not the specimen itself, and so even if the specimen itself is crystalline or periodic, unless these features are resolved in the micrograph, the periodic spacings cannot be observed and analysed by optical diffraction methods.

1.4 Comparison of optical diffraction and electron diffraction techniques

A major advantage of the optical diffraction technique compared with electron diffraction is the small size of specimen area from which it is possible to form a diffraction pattern, with accurate area selection. It was shown in Beeston (§ 5.1.4) that the smallest specimen area that could be selected for diffraction analysis with reasonable accuracy in the conventional electron microscope is about 1 μm in diameter. However, if for example, a micrograph of a selected specimen area is recorded at 100,000 × in the electron microscope, it is possible to select very accurately a 10 mm diameter area of this micrograph for analysis in the optical diffractometer. In this way a 100 nm diameter specimen area may be subjected to diffraction analysis, and of course even smaller specimen areas could be selected by recording the electron micrograph at a higher magnification. In the extreme, it should be possible to analyse a 10 nm diameter specimen area by recording the electron micrograph at 500,000 × and illuminating a 5 mm diameter area of the photographic plate in the optical diffractometer.

In fact, the ability to select a small area is even more clearly an advantage of the optical diffraction technique, since the majority of specimens submitted to optical diffraction analysis contain large periodic spacings generally in excess of 2.5 to 4 nm. Electron diffraction analysis of specimens containing these large periodic spacings necessitates use of low angle diffraction techniques. Accurate specimen area selection is much more difficult in this mode of operation and generally restricts selection to about 40 μm, although 10–15 μm diameter areas may be selected under rigorous conditions (Beeston, § 5.2.2a).

The extension of the optical diffraction technique to optical filtering and image reconstruction is a further major advantage. The procedure is to filter out certain of the diffraction spots or the noise spectrum from the optical diffraction pattern by placing a filtering mask, with holes punched in the appropriate places, in the plane of the diffraction pattern. The mask filters out certain parts of the diffraction pattern and only certain selected beams are transmitted. These beams are then recombined by a further lens system to form an optically filtered image of the original micrograph. In this way it is possible to form separate images from a micrograph of two superimposed periodic images, as with an image of a helical structure, when the front and back of the helix are imaged in superimposition. Alternatively it is possible to filter out the background noise from the image and markedly enhance the contrast and clarity of the periodic information in an electron micrograph. The technique is fully described in Chapter 4. While a fairly similar technique is afforded in the electron microscope by the 'dark field' mode of imaging, the two methods are not exactly analogous and the formation of 'dark field' images using low angle diffraction patterns is not very easily performed.

The optical diffraction technique suffers, however, from a number of limitations in comparison with electron diffraction methods. Not the least of these is the fact that the optical diffraction pattern is formed from the electron micrograph, and therefore is affected by any defects that occur in the formation and recording of the image, such as astigmatism, specimen drift or out-of-focus effects (§ 5.4), and any faults in the photographic emulsion or processing (§ 5.5). Conversely the detection of image and plate defects in this way can be considered to be a useful attribute of the optical diffraction technique. By comparison the electron diffraction pattern is unaffected by specimen drift (provided that the specimen area of interest does not drift out of the selected area), or by image vibration due to mechanical instabilities or external fluctuating fields. However, fluctuations in the accelerating voltage or lens currents may produce blurring effects in the diffraction spots, the effects increasing with the distance of the spots from the centre of the diffraction pattern, and the dimensions of the electron diffraction pattern may be altered by objective lens astigmatism and projector lens distortion, although these are usually fairly minor effects.

A topic of interest to which optical diffraction techniques cannot be applied is the dynamic study of transitions occuring in the electron microscope under the influence of the electron beam, or caused by the application of heat, cold or stress to the specimen. Of course, a retrospective optical diffraction study of the periodic structure may be made from the recorded micrographs, but information cannot be gained sufficiently quickly to influence the course of the experiment in the electron microscope. A subject of particular interest in biological electron microscopy is that of specimen damage by the incident electron beam. While electron diffraction allows an immediate assessment of the state of damage to the crystallinity of the specimen in the electron microscope (Beeston, § 6.2.3), the optical diffraction technique can only report at a later date on the state of crystallinity of the specimen at the moment the micrograph was recorded. This information may, of course, be very valuable, but if any immediate analysis and reaction to that analysis is required, the electron diffraction technique must be used. This is the case in the contrast analysis of 'defects' in metals and alloys when electron diffraction is obligatory in setting up the correct orientation of the specimen with respect to the incident electron beam (Beeston, § 6.1.3).

Finally, as mentioned previously, the optical diffraction technique can only be used to analyse those periodic spacings of the specimen that are recorded on the electron micrograph. By comparison the electron diffraction technique may reveal periodicities that exist in the specimen, many of which may be below the ultimate resolution of the electron microscope.

1.5 Simple optical diffractometer

To construct a simple experimental optical diffractometer it is possible to use inexpensive apparatus normally found in research laboratories. Such a

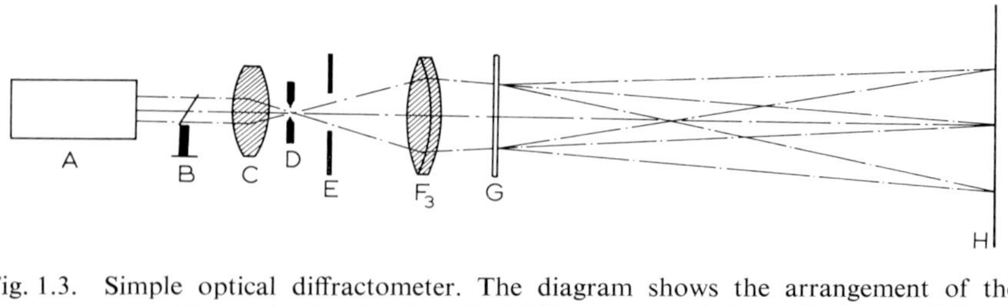

Fig. 1.3. Simple optical diffractometer. The diagram shows the arrangement of the components used to construct a simple optical diffractometer. A. laser; B, shutter; C, beam expanding lens; D, pinhole; E, adjustable diaphragm; F_3, diffraction lens; G, electron micrograph; and H, viewing screen or camera.

diffractometer is illustrated in the diagram shown in Fig. 1.3 and consists of a laser, A, for the light source; a shutter, B; a beam expanding lens, C, which is an 8 mm Beck microscope objective lens; a pinhole, D, which is a 500 μm diameter electron microscope aperture (pinholes of smaller dimensions may be used, down to 30 μm diameter if these are found to be necessary); an adjustable diaphragm, E; and the main diffraction lens, F_3, which can be a simple biconvex lens of about 4 cm diameter and a focal length of about 2 to 3 m, (the choice of this lens is determined by the size of the diffraction pattern required and the length of optical bench available). The electron micrograph, G, is placed after the main diffraction lens, F_3, and the diffraction pattern is viewed on a ground glass screen at H, or recorded photographically.

1.6 Full optical diffractometer for diffraction analysis, optical filtering and image reconstruction

Schematic diagrams of the full optical diffractometer used to extend the technique to optical filtering and image reconstruction are shown in Appendix 1. These diagrams are reproduced at the end of the book for easy reference at any stage in the text.

The diffractometer developed by the authors at the John Innes Institute and described in detail in Chapters 3 and 4 makes use of a long bench, while the Talbot and Polaron diffractometers (§ 7.4) maintain their compact nature by accepting a smaller diffraction pattern.

The full diffractometer system consists of the following components:

- A Two milliwatt helium neon laser, $\lambda = 632.8$ nm.
- B Shutter for photographic exposure of the optical diffraction pattern or reconstructed image.
- C Beam expanding lens.
- D Pinhole to cut out any extraneous light from the laser (sometimes referred to as a side-band filter).
- E Adjustable diaphragm for limiting the illuminated area of the micrograph to about 1–2 cm diameter.
- F_1 and F_2 Collimating lenses to form a parallel beam of light incident on the micrograph. The F_2 lens may be omitted and then the micrograph is placed directly between the F_1 lens and the screen, H, as in the simple diffractometer illustrated in Fig. 1.3. With this arrangement the F_1 lens becomes the main diffraction lens, F_3 (see Appendix 1, lower diagram, b).

G The electron micrograph, or optical grating or test object.

H The diffraction plane, in which may be placed: (a) a ground glass screen for viewing the diffraction pattern.

or (b) a camera system (preferably a Polaroid Land camera back) for recording the diffraction pattern

or (c) a filtering mask, K, in its holder, M; the planes of H and K are shown separated in the diagram for illustration purposes only.

I and J Image reconstruction lens system used to form an image in the plane, N, with the beams allowed to pass through the filtering mask, K.

K Filtering mask.

L Magnification lens to enlarge the reconstructed image to a convenient size for recording. This lens may be omitted if sufficient optical bench space is available to allow a long throw onto the final viewing screen or camera at N.

M Filtering mask holder, made to allow accurate alignment of the mask, K, with the recorded position of the diffraction pattern.

N Final reconstructed image plane, in which may be placed:
(a) a ground glass screen for observing the reconstructed image

or (b) a camera system for recording the reconstructed image.

O Viewing telescope to allow critical adjustment of the optical system so that the reconstructed image is focused on the final viewing screen, at N.

P Magnification lens to enlarge the optical diffraction pattern, if required.

REFERENCES

Bancroft, J. B., G. J. Hills and R. Markham (1967), A study of the self-assembly process in a small spherical virus, Virology *31*, 354.

Bragg, L. (1944), Lightning calculations with light, Nature *154*, 69.

Buerger, M. J. (1941), Optically reciprocal gratings and their application to syntheses of Fourier series, Proc. Natl. Acad. Sci. U.S. *27*, 117.

Castaing, R. and H. Lucien (1964), Filtrage magnétique des vitesses en microscopie électronique, J. Microscopie *3*, 133.

Cosslett, V. E. (1956), Electron microscopy, in:Physical techniques in biological research, Vol. 1., A. W. Oster and G. Pollister, eds. (Academic Press, New York), p. 461.

De Rosier, J. D. (1971), Three-dimensional image reconstruction of helical structures, Phil. Trans. Roy. Soc. Lond. *B261*, 209.

Dowell, W. C. T. (1963), Das elektronenmikroskopische Bild von netzenbenen Scharen und sein Kontrast, Optik *20*, 535.

Fraser, R. D. B. and G. R. Millward (1970), Image averaging by optical filtering, J. Ultrastruct. Res. *31*, 203.

Glaeser, R. M. and G. Thomas (1969), Application of electron diffraction to biological electron microscopy, Biophys. J. *9*, 1073.

Gabor, D. (1965), Information theory in electron microscopy, in: Quantitative electron microscopy, G. F. Bahr and E. H. Zeitler, eds. (Williams & Wilkins, Baltimore), p. 63.

Haine, M. E. and V. E. Cosslett (1961), The electron microscope (E. and F. N. Spon, London).

Hanzen, K. J. (1971), The optical transfer theory of the electron microscope: fundamental principles and applications, in: Adv. Optical and Electron Microscopy, Vol. 4, R. Barer and V. E. Cosslett, eds. (Academic Press, London), p. 1.

Heidenreich, R. D. (1967), Electron phase contrast images of molecular detail, Siemens Rev. *34*, 4.

Klug, A. and D. J. De Rosier (1966), Optical filtering of electron micrographs: reconstruction of one-sided images. Nature *212*, 29.

Lipson, H. and C. A. Taylor (1951), Optical methods in X-ray analysis. II, Fourier transforms and crystal structure determination, Acta Cryst. *4*, 458.

Markham, R. (1968), The optical diffractometer, in: Methods in Virology. Vol. 4, K. Maramorosch and H. Koprowski, eds. (Academic Press, New York), p. 503.

Markham, R., S. Frey and G. J. Hills (1963), Methods for the enhancement of image detail and accentuation of structure in electron microscopy, Virology *20*, 88.

Markham, R., J. H. Hitchborn, G. J. Hills and S. Frey (1964), The anatomy of tobacco mosaic virus, Virology *22*, 342.

McLachlan, D. (1958), Crystal structure and information theory, Proc. Natl. Acad. Sci. U.S. *44*, 948.

Moody, M. F. (1971), Application of optical diffraction to helical structures in bacteriophage tail, Phil. Trans. Roy. Soc. Lond. *B261*, 181.

Reimer, L. (1965), Irradiation changes in organic and inorganic objects, in: Quantitative Electron Microscopy, G. F. Bahr and E. H. Zeitler, eds. (Williams & Wilkins, Baltimore), p. 344.

Ruska, E. (1965), Current efforts to attain the resolution limit of the transmission electron microscope, J. Roy. Microscop. Soc. *84*, 77.

Taylor, C. A. and H. Lipson (1951), Optical methods in crystal structure determination, Nature *167*, 809.

Taylor, C. A. and H. Lipson (1964), Optical transforms (Bell, London).

Thon, F. (1967), On the defocusing dependence of phase contrast in electron microscope images, Siemens Rev. *4*, 13.

Thon, F. (1968), Hochauflösende elektronenmikroskopische Abbildung amorpher Objecte mittels Zweistrahlinterferenzen, in: Fourth European Regional Conference on Electron Microscopy, Vol. 1, D. Steve Bocciarelli, ed. (Tipografia Poliglotta Vaticana, Rome), p. 127.

Unwin, P. N. T. (1971), Phase contrast and interference microscopy with the electron microscope, Phil. Trans. Roy. Soc. Lond. *B261*, 95.

Wilkins, M. F. H., A. R. Stokes, W. E. Seeds and G. Oster (1950), Tobacco mosaic virus crystals and three-dimensional microscopic vision, Nature *166*, 127.

Williams, R. C. and H. W. Fisher (1970a), Disintegration of biological molecules under the electron microscope, Biophys. J. *10*, 53a (Abstract).

Williams, R. C. and H. W. Fisher (1970b), Electron microscopy of tobacco mosaic virus under conditions of minimal beam exposure, J. Mol. Biol. *52*, 121.

Wyckoff, H. W., R. S. Bear, R. S. Morgan and D. Carlstrom (1957), Optical diffractometer for facilitation of X-ray diffraction studies of macromolecular structures, J. Optical Soc. Am. *47*, 1061.

Yada, K. and T. Hibi (1968), Influence of illumination beam on the image contrast, in: Fourth European Conference on Electron Microscopy, Vol. 1, D. Stevé Bocciarelli, ed. (Tipografia Poliglotta Vaticana, Rome), p. 143.

Chapter 2

Light sources

2.1 Introduction

In order for interference to occur between two rays of light they must be coherent with each other. However, light is not normally coherent. Coherence between two rays essentially implies that they left a certain point in space exactly in phase and that they have exactly the same wavelength and plane of polarization. In theory the first requirement is superfluous in that interference merely requires that the phase relationships obtaining when the rays converge should be precise. In practice, however, when using non-laser sources this does necessitate them originating from a point in space at the same instant. In this context a point is a small fraction of a wavelength in size.

2.2 Gas discharge lamps and filtered radiation for optical diffraction

For many years the only artificial source of monochromatic light of any intensity was the gas discharge lamp, suitably equipped with filters. Of such sources the most intense was the mercury arc, which has a wealth of strong lines, almost all of which may be isolated be means of filters. Such light is of course not coherent, in that the rays originate at entirely different points in space and with varying planes of polarization, but a small fraction may be filtered out from the rest having the desired qualities by illuminating a small pinhole with a convergent beam. To do this effectively it is necessary that the Airy disc formed by the beam expanding lens, C (Appendix 1) should be small compared with the pinhole, D. The Airy disc is the central luminous disc together with concentric rings observed when the diffraction pattern formed by a circular aperture illuminated with monochromatic light is imaged in the posterior plane of a lens (Airy 1834). Using a pinhole of 10–20 μm diameter illuminated by the mercury light of the yellow (577 and 579 nm) or green (546 nm) wavelengths, a cone of light can be produced which has sufficient coherence over a solid angle of about 10–20 min to produce adequate diffraction patterns. This light is not completely monochromatic, but light of each wavelength will interfere with itself. With the two mercury yellow lines the two resulting diffraction patterns are so close together that they will appear to be one in any case. However, this is not necessarily so with other types of light source.

The actual amount of light in the beam produced is very small and hardly sufficient for optical diffraction, even using the most intense mercury lamps. Indeed the increase in intensity in the most powerful lamps is usually achieved by raising the arc pressure (and temperature) so that the monochromaticity suffers. A 250 W condensed source lamp is usually employed.

Filtered radiation from sources of this kind has been employed for many years in diffractometers, but only with rather large specimens and very small diffraction patterns. The light itself is so weak that when producing diffraction patterns about 5 mm in diameter from electron micrographs photographic exposures of as much as 18 hr or even more might be required. Also adjustment of the optical system necessitates quite good dark adaptation. Fortunately the low power helium-neon lasers now available provide highly coherent light 10^4–10^5 times the intensity of filtered mercury light and at a capital cost only about 4 times greater.

2.3 Helium–neon laser light source and pinhole systems for optical diffraction

The helium–neon laser provides a narrow beam of polarized red light (632.8 nm) which is almost completely coherent and has a Gaussian beam intensity distribution (Bloom 1963). The red light is accompanied by a small amount of light possessing other wavelengths and by some scattered light.

A perfectly adjusted laser gives completely satisfactory optical diffraction patterns without further filtering, but since the beam is very narrow, typically about 2 mm diameter, a stage of beam expansion, C, is usually incorporated (Appendix 1). The beam may be expanded using either a biconcave lens (Fig. 2.1a) or a biconvex lens (Fig. 2.1b). The former, as in Fig. 2.1a, is more economical of space, and, indeed, is preferable from the point of view of undesirable reflexions from the lens surfaces, which in this case are diluted by divergence. However, the biconvex lens (Fig. 2.1b) has the advantage that the light comes to a true focus, at which a pinhole may be placed so as to re-

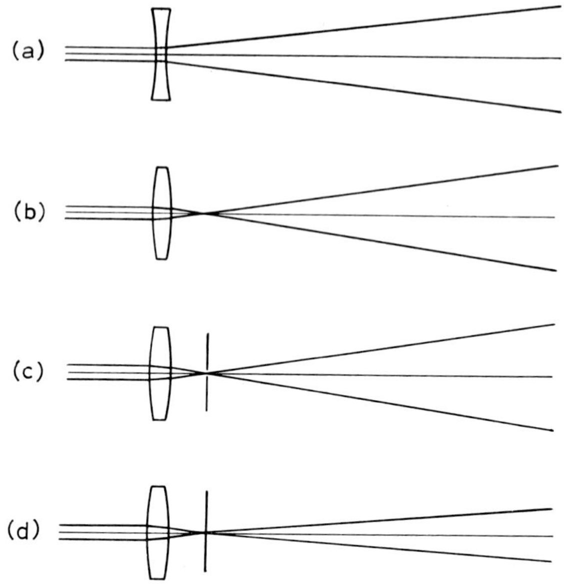

Fig. 2.1. Beam expanding systems. (a) Simple biconcave lens. (b) Simple biconvex lens. (c) Biconvex lens with large pinhole. The spreading of the beam is determined by the focal length of the lens only. (d) Biconvex lens with small (or out of focus) pinhole. The spreading of the beam is controlled by the pinhole diameter. The beam has a central maximum of intensity and a series of concentric cones of decreasing intensity, separated by minima.

move all or most of the stray light, and so produces a perfectly shaped expanding cone of light to illuminate the micrograph (Fig. 2.1c). The size of pinhole required, although much greater than that required for the mercury source, is small (about 100 μm), and since the beam must pass without touching it, its adjustment is fairly critical (see § 3.3). However, this may be done quite adequately using commercially available optical bench fittings.

The actual size of pinhole to be used may be found in the following way. The final beam width at the micrograph should be about 2 cm or more. Indeed there is little point in expanding it less than this because the light intensity available will be found to be ample. If the laser beam has an initial diameter of 2 mm, the beam must be expanded at least 10 times. Suppose the main diffraction lens (see Appendix 1) has a focal distance of 1 m then the beam will need to be expanded using a lens of focal length one tenth this distance, i.e. 10 cm. The Airy disc diameter corresponding to these dimensions is about 80 μm to the first minimum and, since the laser beam has a Gaussian intensity distribution, its Airy pattern only has the central maximum, so a 100 μm pinhole will pass all the laser beam if correctly placed. In practice it may be found desirable to use a smaller focal length beam expanding lens, C, so that a larger pinhole can be used to make the adjustment rather easier. If the pinhole should obstruct the beam at all, it will itself act as a light source and throw out a light beam having a central maximum and subsidiary rings (Fig. 2.1d). The diameter of the central maximum will be inversely proportional to the pinhole diameter. For general use it is advisable not to allow the pinhole to obstruct the beam. It is, however, possible to select a definite beam diameter using a large pinhole, D, placed outside the focus of the beam expanding lens, and then to filter off the subsidiary maxima with an adjustable circular diaphragm (Appendix 1 and Fig. 3.3, E). In this way a circular beam can be shone at the object which can be made smaller than the total micrograph area. With this particular arrangement, since the beam fades away to zero at the edges of the illuminating disc, the diffraction pattern produced contains no information concerning the dimensions of the area of the micrograph being examined, which might clutter up fine detail. A similar result has been achieved by using an oscillating circular diaphragm of variable diameter (Siegel et al. 1966).

2.4 Illuminated area and beam coherence

Having produced an illumination system that has perfect coherence over its whole area, one is bound to ask whether this is the optimal system? Surprisingly the answer is no, so far as the analysis of the electron micrograph is concerned. For studying diffraction from model systems, complete coherence over the whole field is an ideal that in fact the exponents of this art have striven for but have not achieved in most instances, since they have not used lasers. However, what we really want to do in the analysis of electron micrographs is to identify regularities that exist over relatively small distances, and in particular to deduce the structure of the diffracting unit or units. By achieving coherence over a large area we are faced with the fact that such regularities as exist may be diluted by diffraction resulting from the presence of areas not having similar regularities (see for example, Figs. 2.2 and 2.3). This may be overcome be examining only a small selected area of the electron micrograph. However, if the area over which coherence was perfect, was only that of a unit cell or thereabouts the resultant optical diffraction pattern would be the sum of the diffraction patterns over such areas, and so would be exactly what is required, namely a diffraction pattern of the unit cell (see Beeston, § 2.2.1).

(a)

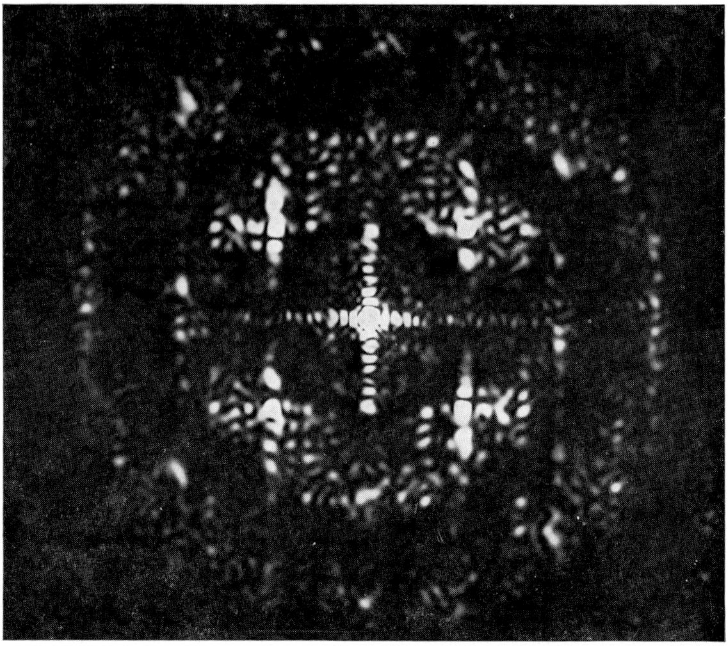

(b)

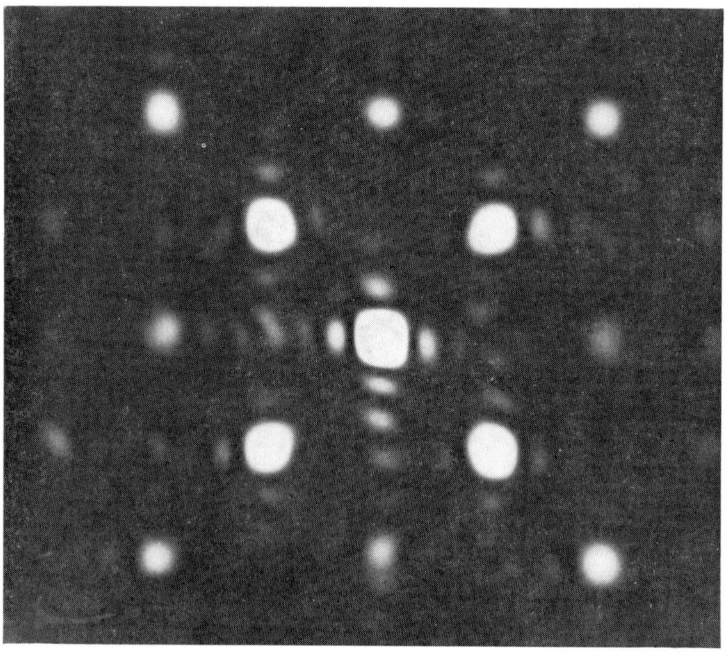

A way in which this dilemma may be resolved is to examine the micrograph over selected small areas in such a way that the coherence of the illumination, in time, rather than in space, is the limitation. To do this a Nipkow-type disc may be used, having adequate line interlace. The Nipkow disc was originally applied as a mechanical optical device for scanning purposes. It consists of an opaque circular disc in which a series of holes are

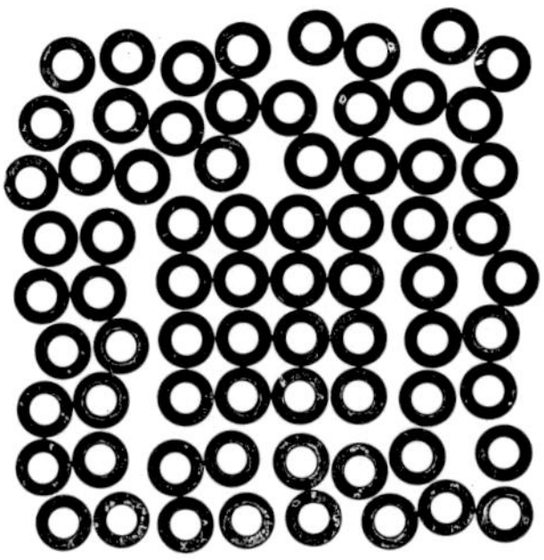

Fig. 2.3. The array of circular rings used in preparing the optical diffraction patterns shown in Fig. 2.2. The actual array was a negative, having light rings on a dark background.

arranged in a spiral configuration about the disc axis. The disc can be illuminated in such a way that when rotated about its axis the light passing through the holes can be projected through an optical system to form a raster or scanning system. The geometry of the holes and their spacing, together with the speed of disc rotation, will determine the number of scanning lines for a given frame interval (Dinsdale 1932). As an alternative, a simple and effective method is to make a small hole, of the size of the area to be selected, in a metal diaphragm and to 'dodge' this by hand over the object, attempting to keep the final uniformity of exposure as even as possible. It is, of course, impossible by these methods to obtain the sum of the diffraction patterns of all the unit cells, but it may well be possible to achieve a much improved resolution of regularities that may exist over restricted small areas on the electron micrograph.

Fig. 2.2. (a) The optical diffraction pattern of an array of rings (Fig. 2.3), part of which is regularly arranged and part of which is randomly arranged. Note that the square arrangement of the regular part can just be identified, but that the pattern corresponding to the diffraction from a single ring is very prominent. (b) The optical diffraction pattern from the ordered part of Fig. 2.3 taken alone. Note the pattern is now much sharper, although the spots are larger because the area examined is smaller. Comparison of the two diffraction patterns shows quite clearly that the presence of an ordered area may be overlooked in a diffraction pattern if a disordered background is also present. Indeed, it may be obscured completely.

REFERENCES

Airy, G. B. (1834), On the diffraction of an object-glass with a circular aperture, Cambridge Phil. Trans. *5*, 283.

Bloom, A. L. (1963), Properties of laser resonators giving uniphase wave fronts, Spectra-Phys. Laser Tech. Bull. *2*, 1.

Dinsdale, A. (1932), First principles of television (Chapman and Hall, London).

Siegel, A., G. J. Hills and R. Markham (1966), *In vitro* and *in vivo* aggregation of the defective PM2 tobacco mosaic virus protein, J. Mol. Biol. *19*, 140.

Chapter 3

Lens requirements and construction of an optical diffractometer for electron microscopy

3.1 General optical and mechanical requirements

In designing a diffractometer that is to be used only for preparing optical diffraction patterns from electron micrographs and not for image recombination, the lenses to be used need only be of a very simple type. In a diffractometer the lenses are always used on the optical axis (Appendix 1), and with monochromatic light, so that the only lens aberration that needs to be considered is spherical aberration. It will be noted that after passing through the micrograph the rays are still effectively on axis and, although the spots will fall on a sphere which is tangential to the plane of the diffraction pattern, H, the radius of this sphere will be so large that the spots will be essentially in focus over a plane surface.

3.1.1 LENSES FOR IMAGING THE DIFFRACTION PATTERN

A simple biconvex lens of 50 cm focal length, used to form a 1:1 image, will be essentially perfect if its aperture does not exceed about 2 cm and will suffice for the main diffraction lens, F_3, because it is unusual for it to be necessary to examine areas recorded on electron micrographs exceeding 2 cm in diameter.

With larger apertures more care needs to be taken. In general spherical aberration increases as the diameter of a lens increases; that is to say scaling up the size of the lens system is disadvantageous. In order to overcome this one may use more complex lens systems. Overall spherical aberration is minimised for simple lens systems by spreading the refraction over as many surfaces as possible. A more advanced lens system is the pair of planoconvex lenses, F_1 and F_2, with their convex faces facing each other. Used as a 1:1 collimating lens system, this combination provides not only approximately the optimum shape for minimising spherical aberration, but it is also distortion free and free from coma, so that this type of lens is also suitable for image reconstruction, where the rays are not confined to the optical axis (see Appendix 1).

3.1.2 ACHROMATIC DOUBLET LENS

A somewhat better lens for the main diffraction lens, F_3, is an achromatic doublet of the type used in small telescopes.

The design of these lenses calls for two glasses with different dispersions and refractive indices, combined as a biconvex lens and a biconcave lens to give a weak positive combination. The flexibility afforded in the design permits the components to be made in such a way that the spherical aberration of one is cancelled by that of the other giving a superlative main diffraction lens, F_3, for a diffractometer. For very large lens apertures achromatic triplets may be used. Of the various types of telescope doublets the non-cemented kinds are better corrected but, because they have four surfaces, they are much more difficult to align precisely. It is our experience, however, that they provide much crisper diffraction patterns than lenses arranged as symmetrical doublets, and are not particularly sensitive to slight inaccuracies in alignment. Note that telescope doublets need to be used with long throws (more than 4 m) or with parallel light and that they are intended for use with their short focal distance on one specific side. This may mean that the ray direction has to be opposite to that when the lens is used as a telescope. Telescope doublets are articles of commerce and may be obtained quite cheaply. Such a lens, with aperture 7.5 cm and focal length 1.25 m, in its mount can be accommodated in a standard optical bench *transverse motion lens holder**. To do this the jaws are removed and the ring is threaded. Details of making this mount are given later in § 3.2.3.

3.1.3 LENSES FOR IMAGE RECONSTRUCTION

As mentioned above a suitable lens system for reconstructing the image is a symmetrical doublet of planoconvex lenses. Using focal lengths for the lenses

* See List of suppliers.

of 4 m, an aperture of about 10 cm may be employed, sufficient to record detail down to about 1.0 nm on the electron micrographs.

Alternatively a telescope doublet can be used followed by a stage of magnification. The lens used for magnification needs to be particularly perfect and a photographic objective is suggested. Unfortunately large aperture telescope objectives are fairly costly, so that one may have to compromise and use a rather small mask for the reconstruction procedure.

3.1.4 LENS SYSTEMS FOR DIFFRACTOMETERS

It will be noted from the above that a multiplicity of lens types may be used in diffractometers. Always provided that length is available in the laboratory it is advisable to use as long focal lengths as possible and simple lens systems. However space may well be a limiting condition. If this should be so then pairs of telescope objectives may be used followed by a stage of magnification. It is even possible to compress the whole system including an image reconstruction stage into quite a small distance by using symmetrical systems of lenses for magnifying and then reducing the images (see Polaron diffractometer, § 7.4.1). Of necessity the lenses would be very expensive and their adjustment would pose some problems.

It will have been noted that the rays passing through the micrograph in some of the systems considered are not parallel. Parallelism of the rays is not in any way obligatory in diffractometers and associated optical systems. Indeed the optical microscope employs a strongly converging ray system in analogous places (e.g. the cone from the illuminating condenser impinging on the specimen). Parallelism of the rays does have a certain advantage in that the diffraction pattern is not influenced in size by changing the position of the micrograph along the optical axis. In non-parallel ray systems the micrograph position has to be recorded, or an internal magnification standard has to be employed. It should be noted that even using parallel rays in the diffractometer section of the apparatus the micrograph position influences the position of the final image in a reconstructing system.

More or less any lens system has its mirror counterpart and diffractometers can be made using mirrors, as well as mirror-lens combinations. Such systems would not appear to have particular advantages to outweigh their obvious disadvantages. The main disadvantage of mirrors is, of course, the necessity for using them off-axis or allowing two passes of the light rays through the object.

3.2 Construction of an optical diffractometer

Diffractometers can be purchased and at present there are at least two types available (see Chapter 7). However, it may well be that particular requirements are better met by constructing one's own system. As an aid in doing this it is possible to purchase optical benches and their accessories as standard items of physical laboratory apparatus*.

3.2.1 LASER MOUNT

The reader is strongly advised to read the instruction manual normally supplied by a laser manufacturer before attempting to place the laser on a mounting device. Most commercially available lasers are supplied with rubber feet for table use and an integral adapter for optional mounting on a standard optical bench. Other makes have various alternative mounting devices. In any event it is essential that the mount should be modified so as to permit the laser tube to be held at a definite height above the optical

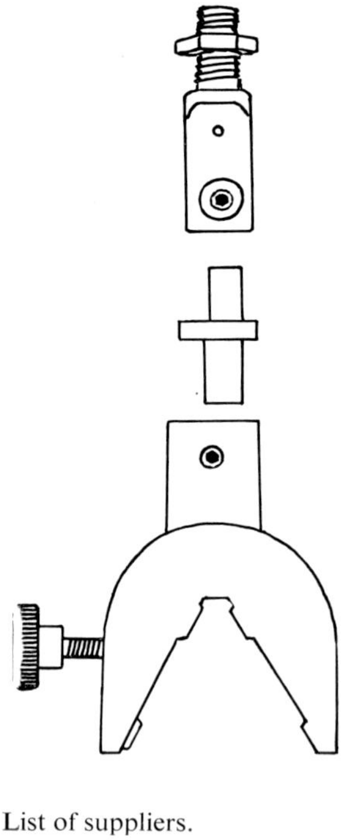

Fig. 3.1. Modified mount for the laser. *Top*: screw fitting made from an adjustable connector from laboratory scaffolding, adapted to fit the laser base screw thread. *Middle*: a cranked pin machined to fit Lablox fittings at the top and the standard mount at the bottom. *Bottom*: standard carrier. The throw of the cranked pin permits side-to-side adjustment of the laser.

* See List of suppliers.

bench, and to be adjustable in all directions (Fig. 3.1), so that the laser beam may be made to lie along the optical bench and parallel to it. In practice the beam itself will define the optical axis of the whole apparatus, so that these adjustments should be precise and easily lock the laser in a permanent position. Alternatively suitable mounts may be fabricated using a *carrier with large flat top** or two of these. On our own new bench, which is mounted on a 5.0 cm^2 hollow steel girder, the mount is made to fit the girder itself. It is equipped with seven adjustment screws (Fig. 3.2) and, once aligned, it need not be touched unless the laser has to be moved or changed.

3.2.2 OPTICAL BENCHES

Optical benches are available in a number of lengths and several designs. The standard triangular (Zeiss) section benches are relatively cheap and have the

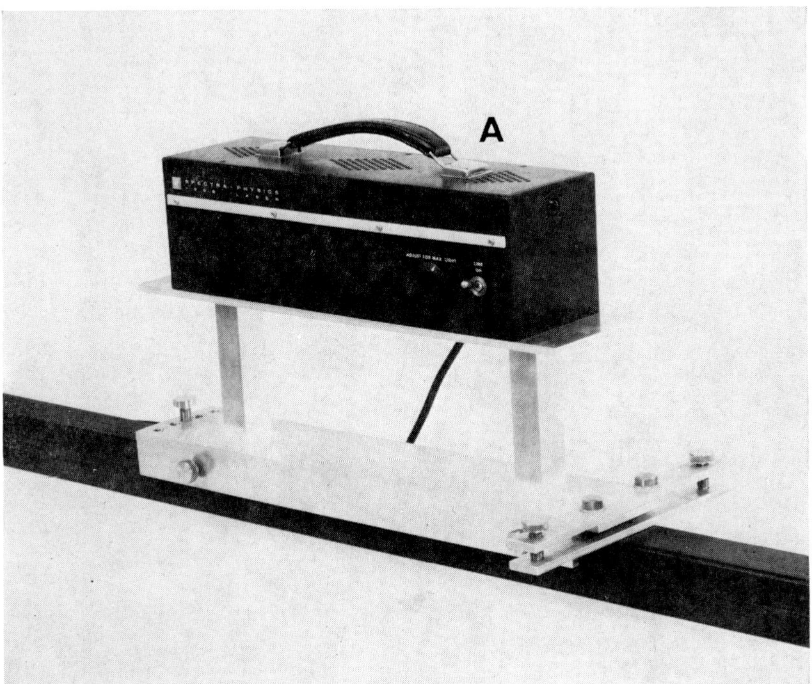

Fig. 3.2. The laser, A, on its mount. Seven adjustment screws are needed for accurate alignment. Five of these (one of which is partially obscured) can be seen in the picture.

* See List of suppliers.

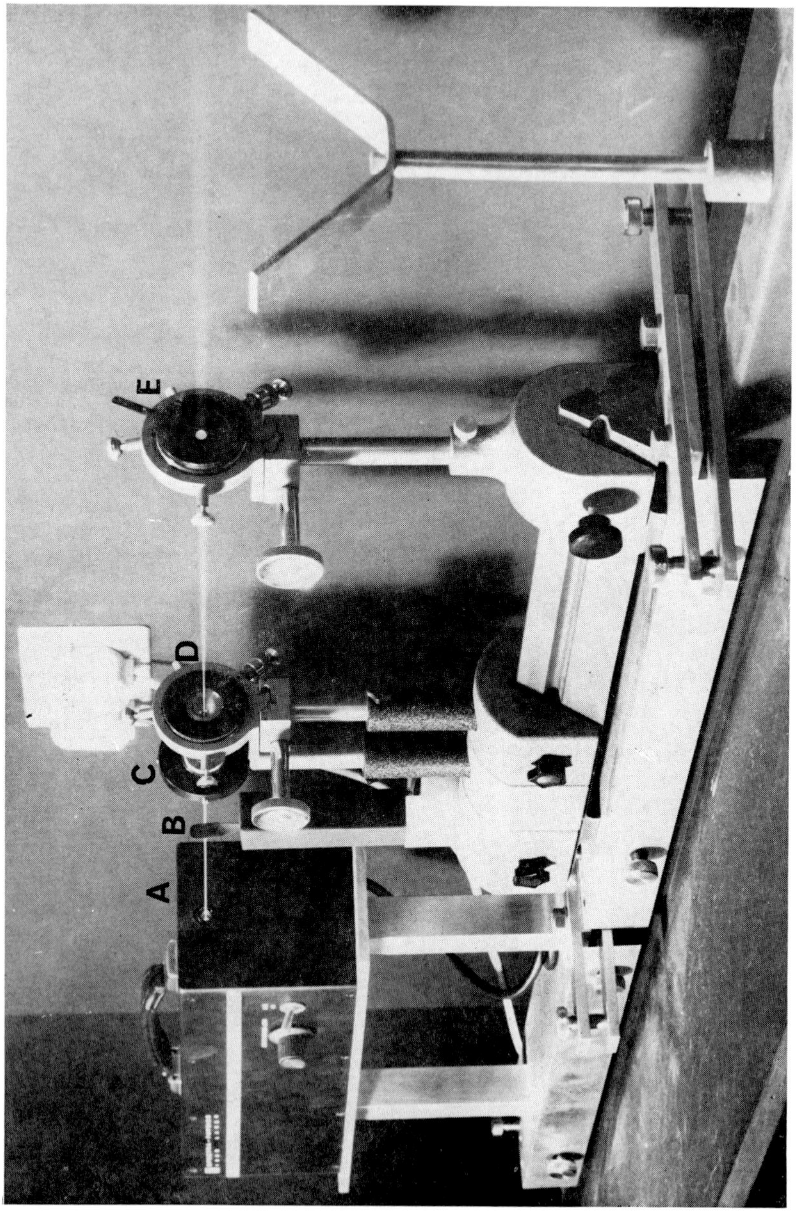

Fig. 3.3. The laser on its mount and the beam expanding system. From the left: the laser, A; the electromagnetic shutter, B; the beam expanding lens, C; the pinhole mount in a focussing, centering holder, D; and an adjustable diaphragm, E. On the extreme right one of the supports for the optical tube used to reduce light scatter from dust particles can be seen. At its base is a button magnet for holding on to the main optical girder. The optical tubes are made from 10 cm diameter plastic drain pipe material.

greatest range of accessories. Most commercial benches of this type are far from accurate in shape, particularly with regard to straightness, and this should be kept in mind when, for example, centering lenses in their mounts. The lens centring may be found to be off quite appreciably if the lens holder should be moved at a later date.

Levelling feet can be obtained for most optical benches, and it will be found economical to get lengths of bench not greater than 1 m and to mount them in line, using some method for clamping the individual sections permanently in position on a laboratory bench. If girder mounting should be contemplated, some sort of adjustment will have to be provided. In our own apparatus, each length of bench is fixed to an inverted U section duralumin carriage which has seven fine screw adjustments for tilt, height and direction. This arrangement (Fig. 3.3) has the great advantage that any bench section may be moved in a matter of minutes to a new location, adjusted precisely and clamped rigidly.

3.2.3 LENS MOUNTS

Small lenses, such as the beam expanding lens, C, may be fitted into a *centring microscope objective holder** or a *focussing and centring microscope objective holder** (Fig. 3.3). Such a mount is also suitable for mounting pinholes.

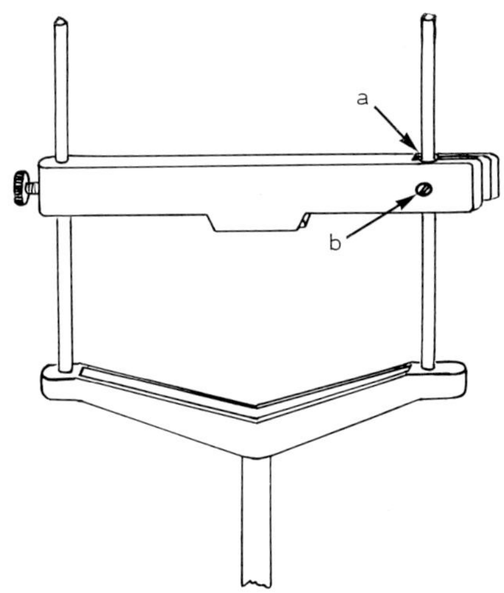

Fig. 3.4. Sliding grip holder modified as described in the text to permit a small amount of tilt to be applied. to the lenses. (a) Indicates the hole coaxial with the vertical support which enables the sliding grip to rotate around the left-hand support. (b) Indicates the position of the small screws enabling the sliding grip to be fixed in the required position.

* See List of suppliers.

The mounting of large lenses presents a serious problem. Optical bench manufacturers make a large number of lens mounts and carriers, but frequently these do not provide any adjustment for forward and backward tilt, and this will have to be provided by the user. So far the best simple solution we have found for this is to use a 10 *cm sliding grip lens holder** modified as in Fig. 3.4. To do the modification take the support and enlarge the end that does not have the clamping screw by drilling, vertically and coaxially with the position of the vertical upright, a hole of about 8 mm diameter. At right angles to this holes are drilled and tapped to take fine screws (6 B.A.) which serve to give the necessary tilt by clamping the upright. This modification only takes minutes and gives a perfect lens mount, which when fitted to a *carrier with vertical travel** and a *transverse travel slide** provides all the

Fig. 3.5. Transverse motion lens holder modified to allow tilt of the main diffraction lens, F_3. The electron micrograph holder and the micrograph, G, are shown at the right of the photograph. One of the plastic optical tubes is seen at the left on its support which is held in position by a small magnet.

* See List of suppliers.

necessary movements. It is also possible to modify the *transverse motion lens** holder (Fig. 3.5) for tilt by removing the top jaw assembly, chamfering the part of the ring beneath and fitting two very fine adjustment screws to enable the top jaw when replaced to rock slightly and then be locked in position.

For mounting telescope achromats, as, for example, the 5 cm diameter, 1.25 m focus lens (obtainable from Brunnings, High Holborn, London) the *transverse lens holder** may be bored out and threaded after removing the jaws and their clamps, or a plate fixed to it into which a suitable thread is cut. Such a mount can then be fitted to a *vertical travel adjustable stem**, which in turn is screwed to a flat plate identical in plan with the top of a *carrier with large flat top**. This carrier has a hole bored in it to allow the downward movement of the *adjustable stem**. The plate is fitted with adjustment screws at the front and back and is pivoted on two steel balls level with the *adjustable stem** and between itself and the carrier.

The image reconstruction doublet, I and J, can be mounted in a length of 12 cm diameter duralumin rod bored to an internal diameter of 9.5 cm and having a coarse thread down its middle to remove internal reflections. The ends are also bored out and threaded to hold screw rings which contain the 10 cm diameter lenses.

In our diffractometer the lens barrel is mounted on a special carrier fabricated from an old carrier that was part of an ultracentrifuge, but a similar device could be made using a *carrier with large flat top** as a basis. The lens barrel itself is held above a 90° trough of duralumin plate which has four adjusting screws, placed so that they contact the barrel radially and their axes, if extended, pass through the optical axis (Fig. 3.6). A spring serves to hold the barrel against these screws. The whole assembly resembles a lathe bell chuck, the upper half of which is imaginary and is 'reflected' by the spring's tension.

3.2.4 CAMERA MOUNTS

Standard 5 in × 4 in camera backs, or other types, can be mounted on standard optical bench mounting pins cut down to the required size (Fig. 3.6).

A special Polaroid film back holder (Fig. 3.6) can be fabricated from brass and mounted on a pillar attached to a *carrier with large flat top**. In order to accommodate filtering masks a dummy of the Polaroid 5 in × 4 in back is made, having slots in it that fit and locate the Polaroid positives. The only

* See List of suppliers.

Fig. 3.6. The image reconstruction lens, I and J, mounted on the optical bench is seen at the right. Note the adjusting screws and spring for lens alignment. A punched mask, K, is shown in position in the mask holder, M, a dummy Polaroid camera back, at the left.

built-in adjustment that needs to be made is a metal strip that locates the bottom of the positives and two screws that serve as vertical adjustments (see Fig. 3.6) to align the filtering mask precisely (see § 4.4.2).

3.2.5 PINHOLES

Pinholes may be purchased made in small discs of metal 3 mm in diameter (Ealing Optical Company) and may be mounted in small mounts to fit *centering microscope objective holders** (Fig. 3.3). They can, of course, also be made by photo-engraving, or by drilling using a very high speed drill or by pushing a thin needle into a thin brass shim. In the latter case the ridge raised on the far side of the shim should be removed with fine emery cloth and then

* See List of suppliers.

the hole reset by pushing the needle in again. The actual size required must be produced by trial and error. The brass may then be blackened chemically. (200 gm copper carbonate dissolved in 700 ml 0.880 S.G. ammonia. Dilute to 1200 ml. Degrease the brass and immerse until coloured adequately and finally rinse in water).

3.3 The alignment procedure

The laser on its stand is set up at one end of the girder and aligned about vertical and horizontal axes using the levelling screws. The beam is adjusted parallel with the girder and vertically above its centre.

Important: Never look directly at the laser beam.

A section of optical bench is then set up as near to the laser as possible and levelled. A lens carrier with a centring pin (Fig. 3.7) is set up with the tip of the pin at the height of the centre of the laser beam. The optical bench is then adjusted laterally so that the laser beam falls on the tip of the pin when the pin is placed at either end of the bench. Then a second section of optical bench is set up similarly at the farthest end of the girder and adjusted in the same manner. Note that the centring pin is now a secondary standard. It must not be adjusted in any way, and should be locked to its carrier by a grub screw.

The laser beam of light now defines the optical axis of the diffractometer. It is useful to put white disc-shaped targets at each end of the bench, or on the walls if conveniently situated, to define the position of this light beam permanently, so that lasers can be changed without upsetting the position of the optical axis.

Any intermediate section of bench may now be put in position using the pin and the laser beam for reference. In doing this it is also necessary to check the level of each section.

The main lenses are now put into position one at a time. The laser beam strikes the lens surfaces and some light is reflected back. The reflected beams are centred on the laser exit window. The more important adjustments are for tilt and rotation about a vertical axis. The fine vertical and lateral adjustments may be checked by ensuring that the laser beam still strikes the white target at the far end of the bench.

The main diffraction lens, F_3, or lenses, F_1 and F_2, are now put into their final position. Then the beam expanding lens, C, is set up near the laser, A, adjusted for position and finally adjusted by centring the expanded beam onto the axis of the main diffraction lens. Having done this *no further* adjust-

Fig. 3.7. The pin used in setting up the optical axis of the diffractometer. Note the shadow of the point on the card, K, which is held in the mask holder, a dummy Polaroid camera back. Behind this is the reconstruction lens system, and in the foreground can be seen the control switch for the shutter which has both toggle and push-button actions.

Chapter 4

Image reconstruction

4.1 Introduction

Image reconstruction techniques have been used with remarkable effect in normal photography especially to restore focus or to change phase. In this type of exercise it is not usual to subtract detail from the diffraction patterns but to modify them by changing the amplitude and the phase of zones of the patterns. The procedure is, of course, very demanding in technique, particularly because diffraction effects are not additive, so that it is necessary to make graded amplitude filters and phase filters to obtain the required result.

From these considerations it will be seen that the mathematical reconstruction of an object from its recorded optical diffraction pattern is a matter of some difficulty (Chapter 9). However, if a reconstruction lens is put at or near the plane of the diffraction pattern and focused onto the micrograph to form an image on the posterior side (Appendix 1b), the reconstruction follows automatically, provided that all the diffraction spots are within the aperture of this lens and, of course, that the lens is free of aberrations.

In many optical systems the diffraction pattern is either inaccessible physically or it is not sufficiently perfectly resolved to permit modification other than by interposing crude stops or diaphragms, but in a well designed optical diffractometer the diffraction pattern is not only accessible but may be several centimetres in diameter. It is now easily modified in such a way as to influence the final image and, indeed, has been used quite successfully for this purpose.

4.2 Spatial frequency or optical filtering

Both Klug and De Rosier (1966) and Bancroft, Hills and Markham (1967) independently used the optical diffractometer for the purposes of optical filtering or spatial frequency filtering, as this process is generally called, to record obscured detail originating from one side of helical objects. To what extent is such manipulation justifiable? Klug and De Rosier have argued that since the major spots in the diffraction patterns of helical objects originate from opposite sides of the helices, i.e. there is zero optical interaction between the two sides, the process enables the dissection to be carried out reasonably satisfactorily. Within the limits of their argument there is no question that dissection is possible, provided that the helix is not restricted by a filtering mask, a qualification that can be met, for example, by using a predominantly light helix on a black background. However, as soon as the helix becomes distorted by flattening, or worse by internal disorganisation, optical interaction between the two sides occurs. This interaction has its origin in the very nature of optical diffraction and while the resulting effects may be trivial this is not so in many instances.

Optical diffraction patterns contain spots originating from rays diffracted both by structures in the micrograph G, and in the mask, used to select an area of the micrograph, if any. These structures are rarely, if ever, precisely regular and free from background noise. In consequence, the optical diffraction pattern is produced by a complex interaction process in which every individual modulation in the object takes part. In its nature therefore, the diffraction pattern is a product rather than a sum and so except in the case of a very restricted set of diffraction spots of specific objects, simple subtraction of individual (pairs of) diffraction spots, has effects very different from those anticipated. This concept may be rather difficult to visualise without concrete examples. In a regular crystalline array of particles, or in a helix having a regular substructure, the number of distances between pairs of particles, and their relative orientations is limited. If they are randomised even slightly from these specific positions, both their relative separations and orientations are affected and so they have a very much greater number of possibilities of interacting in diffraction. This is illustrated in Figs. 4.1 and 4.2. Even with most crystalline arrays, the diffraction pattern is in any case by no means a summation of the diffraction patterns of its components (Fig. 4.4) so that the subtraction of spots does not have the desired result of only removing certain detail selectively. It is also well established in optics that interference with diffraction spots can be a possible source of artefact (Figs. 4.3a and b).

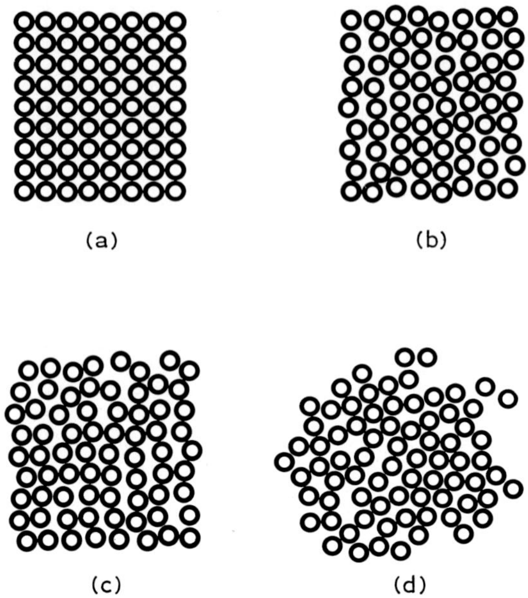

Fig. 4.1 a, b, c, d. Arrays of circular rings ranging from perfect square alignment (a) to complete disorder (d). Optical diffraction patterns made from negative copies of these arrays are shown in Fig. 4.2.

The net effect of these factors is somewhat discouraging. It would appear that except in the case of mathematically perfect objects having certain symmetry features, optical filtering by the absolute *subtraction* of spots, can only result in loss of overall resolution and is liable to produce (predictable) artefact structures. However, if a loss in resolution is acceptable, the technique has it uses, particularly in unscrambling Moiré type structures (Fig. 4.5). As a general guide, the method is most successful with perfectly preserved structures, particularly helices (with the reservation that some minor spots may overlap). Pairs of hexagonal lattices have diffraction spots that interact less than most others, provided that the angular separation between the lattices is kept small, and indeed such objects greatly resemble helical objects which in any case usually collapse to a structure intermediate in form between a true helix and a pair of flat hexagonal lattices.

It goes without saying that when two lattices overlap in the way discussed above parts of each lattice, especially the lower one, are completely obscured in the electron microscope image. Generally, however, the reconstructed image does not show this, the diffraction procedure tending to fill-in the gaps with a

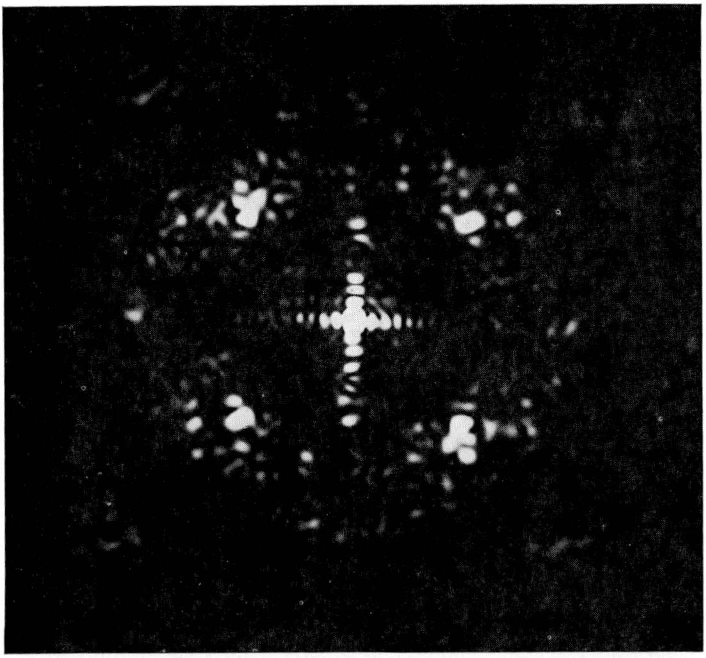

(b)

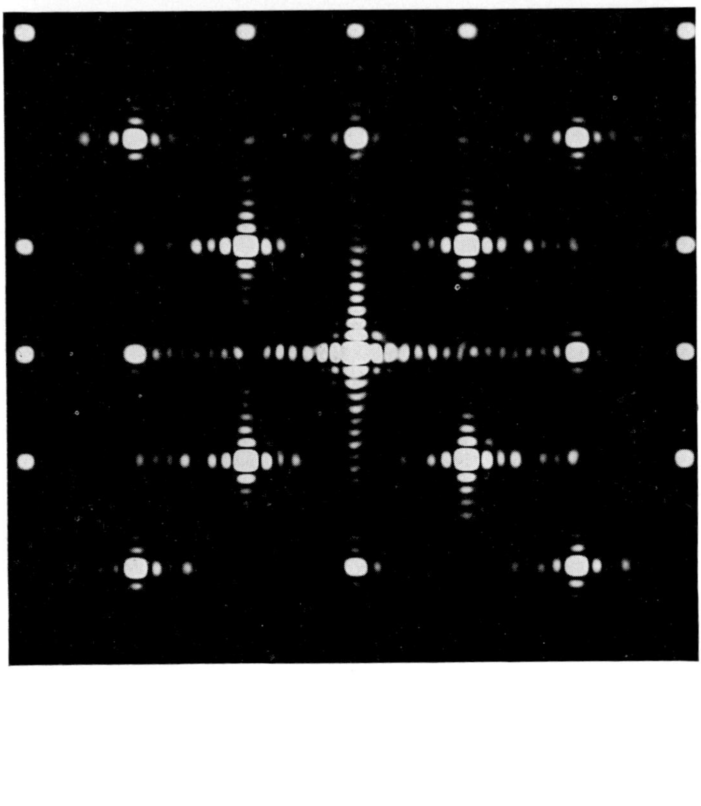

(a)

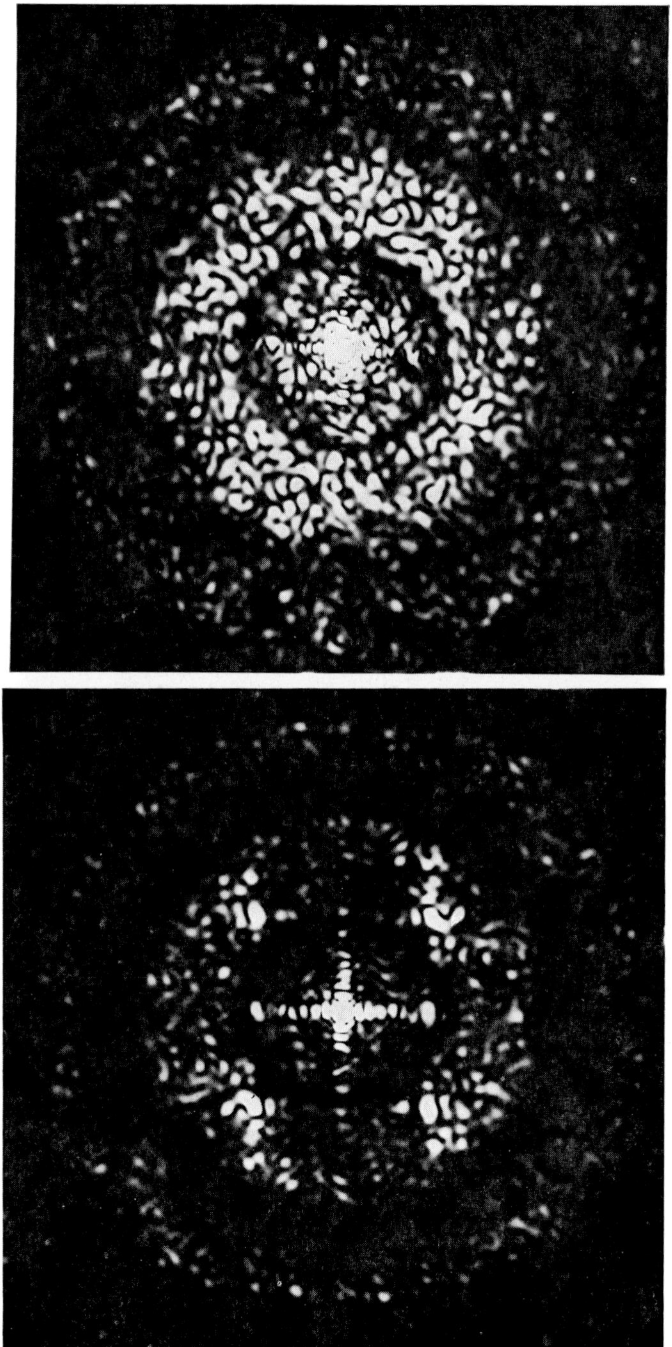

(c) (d)

Fig. 4.2 a, b, c, d. Optical diffraction patterns made from negative copies of the arrays of rings in Fig. 4.1 a, b, c, d, respectively. Note that the diffraction pattern of the regular square array (a) gradually disappears and is replaced by an approximation to the diffraction pattern for a single ring (d). Note also that the first order diffraction spots in (a) fall within the limits of the rings in (d).

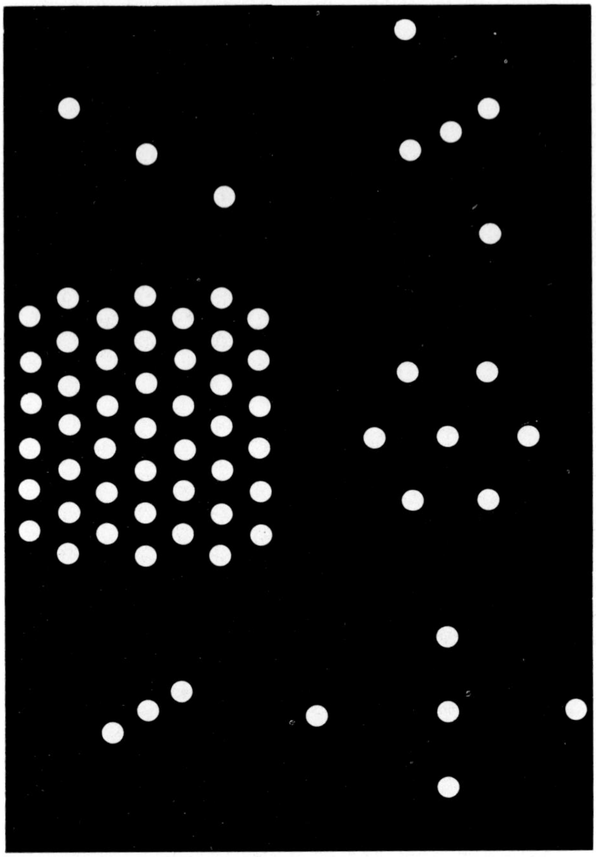

Fig. 4.3a. The diffraction pattern of a hexagonal mesh is shown (middle left), together with five different filtering masks. The reconstructed images formed using the diffraction spots selected by these masks are shown in Fig. 4.3b.

structure programmed by the filtering mask. This peculiarity is useful to demonstrate the substructure of helices formed from subunits (Figs. 4.6 and 4.7).

4.3 Optical filtering by attenuation of the central spot of the diffraction pattern

A simple and undemanding application of optical filtering methods, which so far has not been used in electron microscopy, is to accentuate the fine details in the image by attenuating the central spot of the diffraction pattern.

Fig. 4.3b. Images formed by optical filtering and reconstruction using selected spots in the diffraction pattern of a hexagonal mesh (see Fig. 4.3a). The image shown on the middle right is particularly interesting because not only is the hexagonal lattice rotated by 30° but bright spots also occur where there was nothing in the original specimen. It will be noted, particularly at top and bottom right, that the apparent detail in the final images is greater than one would expect from the effective apertures used. This is a phenomenon caused by the rapid cut-off imposed by the diaphragms. Although the examples shown are rather extreme, similar artefacts are likely to occur whenever one interferes with the diffraction pattern.

All that is required is a series of graded filters having maximum absorption in the centre. Such a filter can be made by sandwiching a photographic area, having a maximum density in the centre and a bell-shaped density distribution, between optical flats and placing it in the plane in which the diffraction pattern is formed. This may be made by photographing the expanded laser

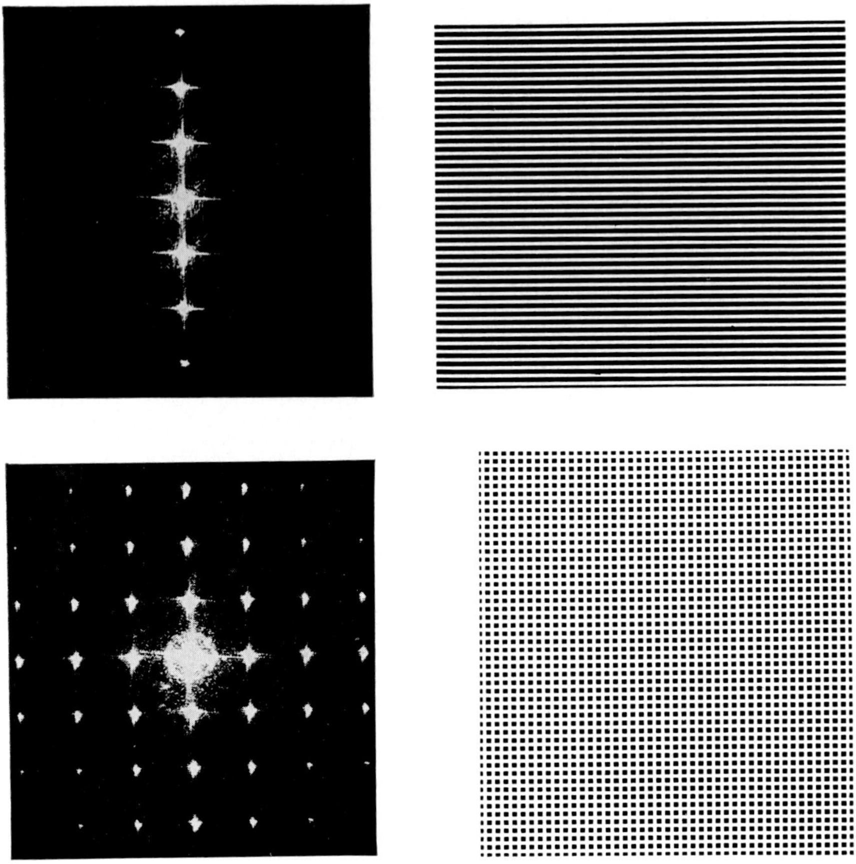

Fig. 4.4. Model diffraction patterns showing that the optical diffraction pattern of two superimposed objects is not the sum of the diffraction patterns of the individual objects and therefore that it is not in general possible to dissect out detail of one from the complex by the simple application of optical filtering methods. *Top left*: diffraction pattern of a line grating; grating on right. *Bottom left*: diffraction pattern of two crossed identical gratings; object on right. It is interesting to note that if the diffraction pattern contained in the top illustration were subtracted from the bottom diffraction pattern with the aid of image reconstruction methods, a rectangular grid image would still be formed and not a vertical line grating.

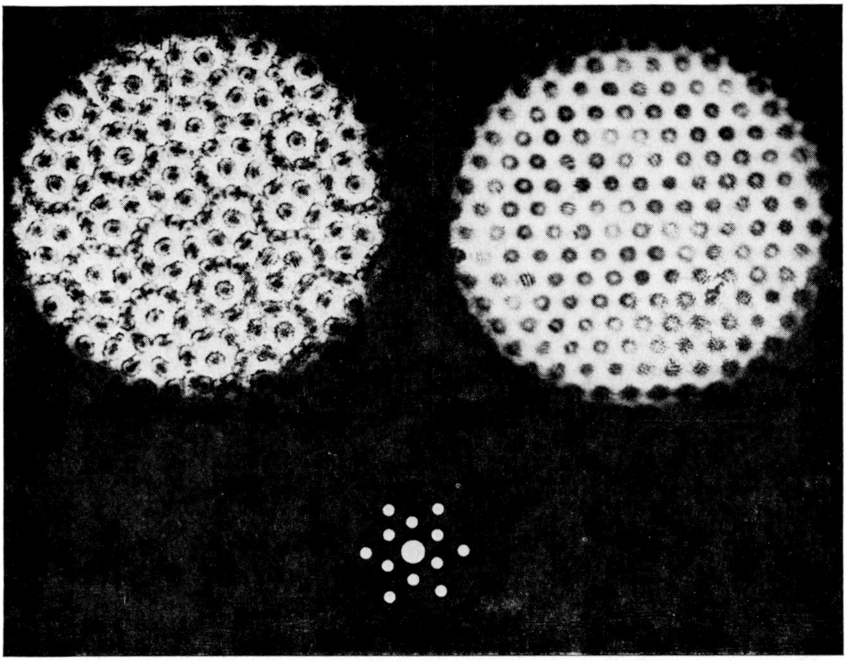

Fig. 4.5. The unscrambling of a Moiré pattern. Left, the pattern formed by the superimposing of two identical hexagonal dot patterns not in register. Right, the reconstructed image of one pattern, using the simple filter shown at the bottom of the diagram.

beam directly on film using a diffusing screen near the film and an attenuator (dark filter), to reduce the beam intensity. It should be noted that mesh filters, as used by Klug and De Rosier (1966) for attenuating the central beam, are in fact not suitable in this precise position unless the diffraction spots should happen to be extremely ill-defined, because mesh filters cut off sharply defined areas completely, rather than attenuating partially over relatively large areas.

4.4 Image reconstruction techniques

The procedures involved in reconstructing optically filtered images are difficult technically, but the degree of difficulty depends very much upon the overall design of the optical diffractometer being used. The problem is that one must provide a filtering mask, opaque except for those areas through which the required diffraction beams must pass, and it must be accurately located in the

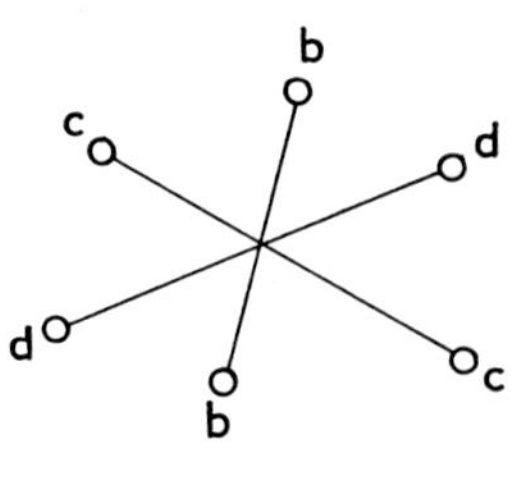

Fig. 4.6. (a) drawing of the projection of one side of a helix made up from hexagonally arranged subunits and, below, its optical diffraction pattern. The resolution is limited in the drawing so as to give only one order of diffraction. (b), (c) and (d) are reconstructed images of (a) each using one pair of diffracted spots, similarly labelled, plus the central carrier. It will be noted that whole helices are generated, one in (b) two in (c) and three in (d). These correspond to the main ways in which the subunits in (a) are arranged, but since the information afforded for true reconstruction is inadequate the absent 'detail' is filled in automatically.

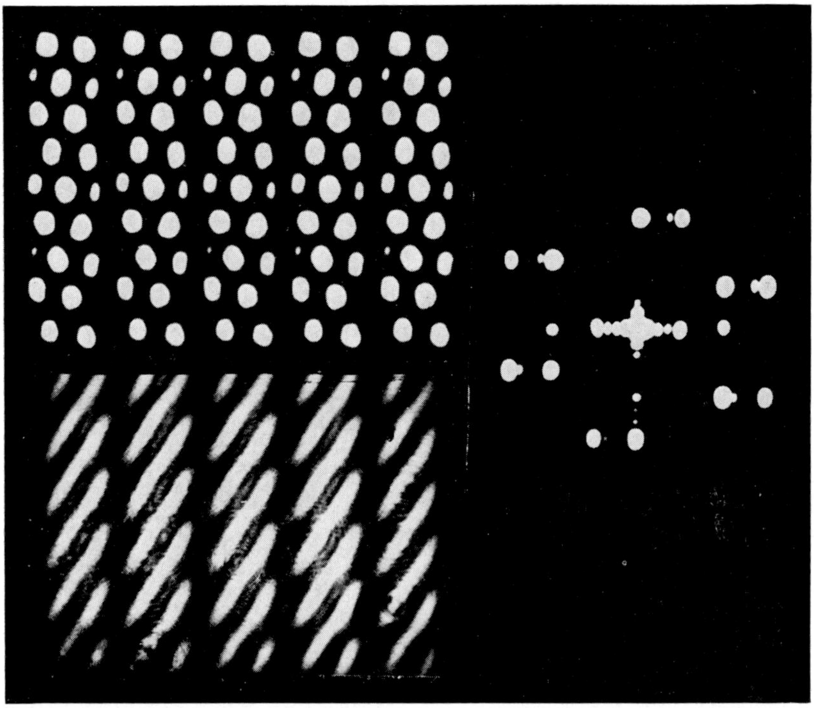

Fig. 4.7. *Top left*: an array of five helices, each similar to the one in Fig. 4.6. *Below left*: the array reconstituted using the diffraction spots corresponding to (c) in Fig. 4.6. *Right*: the diffraction pattern of the array. Note now that the spots are smaller but still fall within the areas of the spots from one individual helix. Note also that the horizontal spacing between major spots now contains the information corresponding to the distance between adjacent helices. The fine horizontal modulation that can also be seen represents the distance between the first and the fifth helix and so gives the overall width of the array. The fine vertical modulation represents the height of the array and the shape of the larger spots indicates the overall dimensions of the array.

plane in which the diffraction pattern is formed and oriented precisely with the diffraction pattern. This has been accomplished in the design of the John Innes diffractometer in such a way as to permit the whole process to be carried out in a matter of minutes. This procedure will be described first.

4.4.1 MAKING THE FILTERING MASK

The diffractometer is provided with a special holder (Appendix 1, M) which is just a skeleton frame designed to hold a Polaroid 5 in × 4 in camera back, and this is placed in the plane in which the diffraction pattern is formed (H). If the original electron micrograph plate magnification should be incorrect

(§ 5.6), a positive print is made on film at the appropriate magnification. A photograph of the diffraction pattern is then taken on Polaroid P/N film (§ 3.5), and the positions of the required spots or areas are marked out on the positive. Holes surrounding these spots or areas are punched out with a leather punch or cut out with a fine scalpel, or pricked out with a red-hot needle. The punched or perforated positive provides the filtering mask in the form of a piece of thin card (Fig. 3.6, K). Leather punches have rather narrow throats and a longitudinal slit in the photograph will assist punching the holes and the slit can be taped together later. Naturally such slits should avoid important detail in the diffraction pattern.

4.4.2 REGISTERING THE FILTERING MASK

Polaroid 5 in × 4 in P/N film is made in such a way that the lateral register of the film and the card positive is excellent with respect to the film back and holder. The vertical register is probably just as good in the unexposed film package, but since the positive card is withdrawn during exposure it may not be replaced precisely in its vertical position before the image transfer from the film to the card takes place. Consequently the holder, M, made to take the mask K, needs to have a small amount of vertical adjustment (about ± 2 mm) but no lateral movement is usually necessary. If lateral adjustment should be required it may be provided by means of the optical bench adjustment screws.

The filtering mask holder itself (Fig. 3.6, M) is essentially a sheath made to hold the positive and having the same overall dimensions as the appropriate parts of the Polaroid film back. All that is usually necessary to register the mask with the diffraction pattern is to check that the vertical adjustment is correct on the positive print. If not, the vertical movement screws are adjusted to compensate. Then the mask is in complete register. This register may be checked by holding a piece of translucent film behind the mask and seeing that all the required spots appear on it.

4.4.3 PRODUCTION OF MASKS FOR SMALL DIFFRACTION PATTERNS

Making masks for smaller diffraction patterns is more difficult. One method that we have used is to put the negative picture of the diffraction pattern in an enlarger and enlarge it to a workable size. The areas required to be punched out are then marked out on a sheet of tracing paper placed over the enlarged pattern, which is put on the enlarger easel and illuminated. A lantern plate or film is put in the enlarger plate carrier, and the enlarger itself is used as a camera to reduce the filtering mask back to the exact size. This plate or a contact print can then be used as the filtering mask, or one or other as appropriate can be used to produce a photo-engraving on brass shim-stock by ordinary photo-engraving methods. This latter procedure, although tedious, is the preferred one. Small filtering masks may also be made using a pantograph-controlled drill or punch*.

Registration of the small mask with the diffraction pattern presents problems. The mask must be held in a holder capable of being moved vertically, and horizontally, and of rotating in a plane at right angles to the optical axis. Adjustment is facilitated by using a low power viewing microscope on an optical bench carrier to check that the diffraction spots lie inside the filtering mask holes, in which event the objective may be changed for a low-power telescope objective to look at the specimen itself through the mask**. Having centred the mask on the centre of the diffraction pattern, it is then rotated to allow the other spots to pass. Often it is impossible to see them clearly, in which case two objects must be examined at one and the same time. A picture of a grating or of a suitable periodic object (see § 6.5) of any kind with a very fine structure is mounted in the same plane as the object to be examined. When taking the original picture of the diffraction pattern, two exposures are made on the same film, one with the main object and one with the grating, the other being masked off as appropriate. The fine grating will give strong spots spaced out sufficiently to enable registration to be carried out with ease. When this has been done the other parts of the filtering mask are automatically in register. A very similar procedure is possible with electron micrographs copied from published half-tone pictures which will already have information for the half-tone screen spots on them.

A refinement of the above procedure has also been used to remove certain specific spots while passing everything else. In the large diffractometer this is easy to do. All one has to do is to put an optical flat in the diffraction plane (note the defocusing of the reconstructed image caused by the optical flat), and black out the appropriate areas with ink on the optical flat. With small diffraction patterns, it is necessary to make a photo-engraved filtering mask and then to make the necessary opaque spots on a clear background by using vacuum evaporation of aluminium onto a small optical flat through the mask.

* See List of suppliers.

** The same precautions concerning viewing the laser beam apply here also.

4.4.4 DESIGNING THE FILTERING MASK

In the preceding section, the crucial question of which spots are to be transmitted by the filtering mask has been very conveniently omitted. However, the question still remains. In some specific cases the answer is not particularly difficult. In a one dimensional diffraction pattern, the spots or lines are at distances from the central spot which bear a simple linear relationship to each other. Some spots may be missing or suppressed, but their location is quite obvious.

In a two dimensional pattern the same general rules obtain, but in this case one may draw a number of straight lines passing through certain rows of spots. These lines intersect to form a definite and relatively simple pattern. If two lattices are present one can usually deduce which is which by this elementary geometrical approach (Fig. 5.1). Note, however, that is it only too easy to fail to realise where the appropriate spots should lie if they should be missing or even if they are sufficiently strong to be seen, but just to weak to notice!

The simple geometrical relationship between the positions of the spots enables one to employ the convention, usual in X-ray crystallography, of *layer-lines* (Lipson and Lipson 1969) and row lines, which is of definite utility and gives one a series of co-ordinates to which one can refer the spots (Fig. 4.8). One cannot help but note however, that in one of the earliest published applications of diffraction to a two sided electron micrograph ill-defined, but quite evident, major spots, corresponding with the *main* lattice periodicity of the object would correspond to a 'layer-line' number of $2\frac{1}{3}$! In this particular case the structure, deduced at great length, was misoriented by 30°, and, of course, the dimensions found were also incorrect.

One of the major difficulties in allocating co-ordinates to spots is that very often it is impossible *a priori* to decide which of a number of possible spots is the correct one. Quite minor disarray in the specimen may result in the production of multiple spots. And very often in specimens known to have two stained sides, the periodicities on back and front are quite significantly discordant. Also, since what one usually sees is the distribution of stain in the specimen, the penetration in depth of the stain may be, and usually is, different on both sides. This is particularly true in the case of helical objects, which have the further pitfall that the spots on the transform of a helix are not discrete but have satellite maxima corresponding with the successive maxima of Bessel functions (see Lipson and Lipson 1969) (Fig. 4.8). While these are quite easy to pick out, they have the inconvenient property of sharing the same position in reciprocal space with spots from the other side

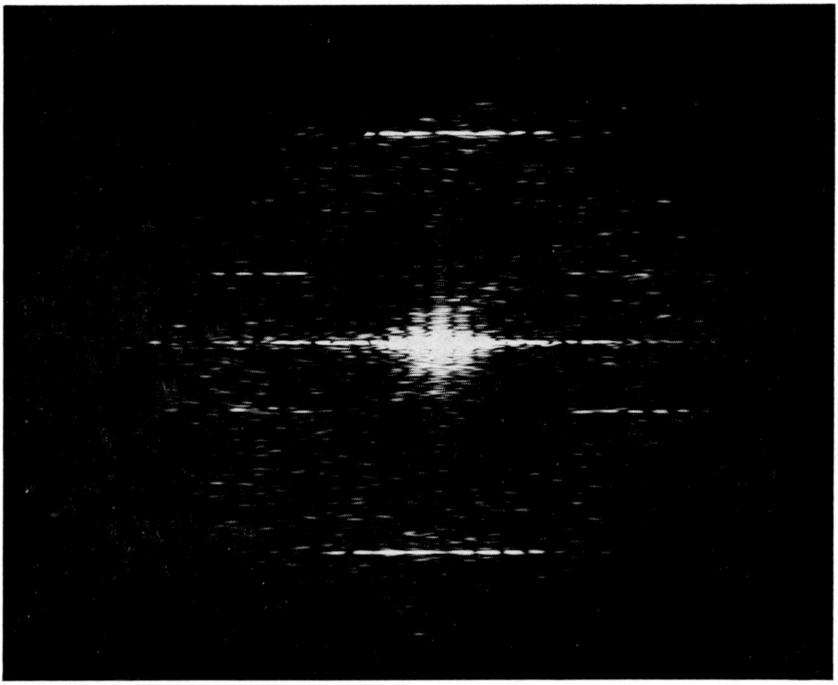

Fig. 4.8. Diffraction pattern of a single tobacco mosaic virus particle taken using a simple plano-convex lens of 3 m focal length. The axis of the rod is vertical, and the top and bottom (3rd layer) lines correspond with the short helix pitch (2.3 nm). The first layer lines are also well resolved and the second can just be seen. The particle used was one of the particles in the field shown in Fig. 1.2.

of the helix, a factor that militates against the clear subtraction of the diffraction spots by simple mechanical manipulations.

4.5 Designing the reconstruction lens system

The image reconstruction lens I and J (see Appendix 1) needs to be of long focal length and to have a large aperture (up to 9 cm diameter, the width of Polaroid film). It also needs to give an undistorted image. There are two possible choices for this lens. One is an astronomical objective, which will need a supplementary stage of magnification (L), or a specially made distortionless (rectilinear) lens. The system shown in Figs. 3.6 and 3.7 consists of two 10 cm diameter plano-convex lenses with their convex sides facing each other. The focal length of each of these lenses is 4 m and the combined lens is used to give an approximately 1:1 image of the object being examined. A final × 4 magnification may be provided by a 35 mm photomicrographic aplanat, L, but in this position any lens of the types used in photographic enlargers may be used. Photomicrographic lenses have the advantage that having R.M.S.* standard threads they fit standard optical bench lens holders. Generally speaking this stage of magnification should be avoided if possible because it complicates the adjustment of the apparatus and may introduce aberrations.

4.5.1 FOCUSING THE RECONSTRUCTION LENS SYSTEM

Having provided the appropriate lenses and filtering mask all that now remains is to focus the reconstructed image using a viewing screen or viewing microscope, O, and photograph it. However, this is about the most difficult thing to do. It is quite useless to attempt to focus the image itself with or without the mask in position, because it is almost impossible to distinguish the true image from false but equally, or even more, plausible images. A large number of focusing test objects may be used, including transparencies of typewriting much reduced, but possibly the most suitable object is a single pinhole about 50 μm in diameter in the centre of the field. It goes without saying that the focusing of the recombining lens system must be carried out before the filtering mask is made. It may in fact be done once for all time provided that the micrograph holder is not moved subsequently.

A procedure that may be used for focusing the reconstructed image precisely is as follows. The whole apparatus is aligned as well as possible, in particular the micrograph holder, G, which is then fixed in position. The main diffraction lens, F_3, is removed. A 13 cm lens in a *centring and focusing microscope objective* holder (see List of suppliers) is placed in position and used to form an image of the pinhole, D, in the plane of the micrograph, which is left in position. This image is then used as the pinhole source for a diffractometer which uses the reconstruction lens as its main lens. The diffraction pattern from this diffractometer is then formed in the position of exact focus of the reconstruction lens and is observed by means of a *pin-mounted*** viewing microscope or an *optical bench saddle***, mounted on a section of optical bench at the requisite distance (Fig. 4.9). All intermediate filters, etc. if any, should be in position before final focusing is carried out.

The image seen in the viewing microscope is an Airy disc with a bright centre. Out of focus either way the centre becomes dark and then light and

* Royal Microscopial Society.
** See List of suppliers.

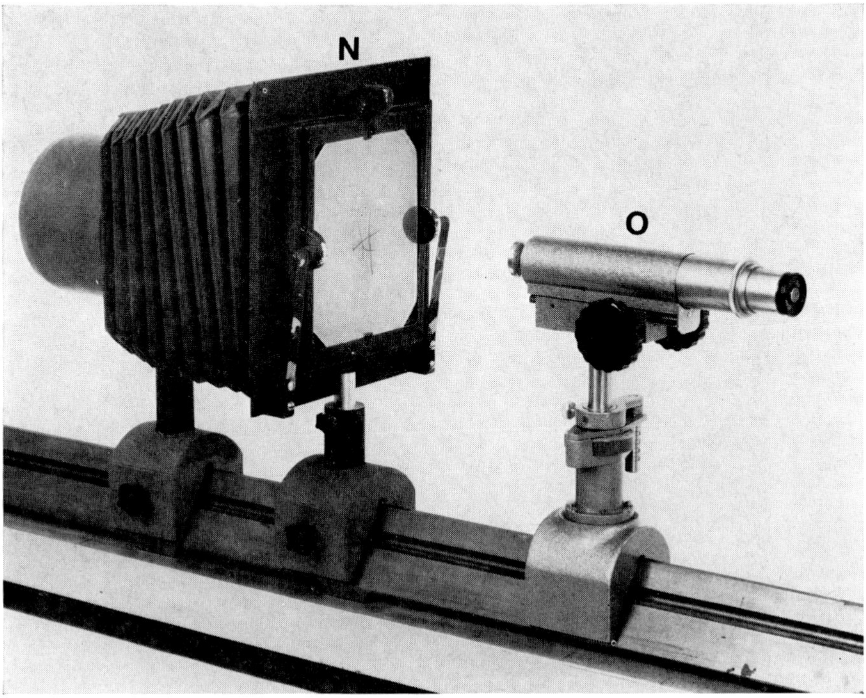

Fig. 4.9. Viewing microscope, O, being used to ensure that the test diffraction pattern and the ground glass screen of the camera are in precise register. This is the final stage in focusing the image reconstruction lens system, I and J. (See Appendix 1.)

so on as the focus position is changed. The first dark positions on either side can be observed and the distance between them bisected.

An alternative and much simpler method is to obscure the aperture of the reconstruction lens with a card with two circular holes placed as far apart on the aperture as possible. The diffraction pattern of a pair of holes is a series of concentric rings the diameter of which is determined by the inverse of the hole diameters, crossed by a series of parallel lines spaced inversely proportionally to the distance between the hole centres and at right angles to the line joining the hole centres (Fig. 4.10). Incidentally, the distance between these parallel lines is the minimum distance that can be resolved by the reconstruction lens if its total aperture is equal to the distance between the hole centres.

The John Innes diffractometer reconstruction lens has a clear aperture

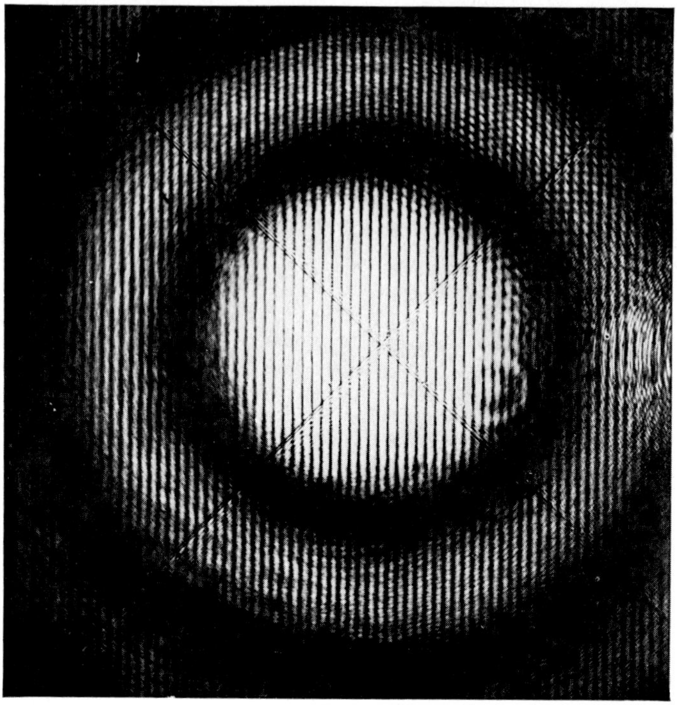

Fig. 4.10. Image of the Airy discs from two holes modulated by the distance between them, as seen through the viewing microscope when the reconstruction lens is precisely in focus. The distance between the vertical lines is the smallest detail that can be resolved in the reconstructed image since it represents the diffraction limited aperture for the reconstruction lens.

of 9.5 cm, so the two holes, which are punched on a card that is placed in front of the lens are 7–8 cm apart and are 6–8 mm diameter.

When the diffraction pattern is observed and is out of focus, two sets of concentric rings will be seen. The viewing microscope is moved so that these overlap and fuse and the fine parallel lines are sharply defined (Fig. 4.10). A camera back holding a ground glass screen is moved into position so that the ground glass is also in focus in the microscope. The diffraction pattern is then in focus on the ground glass screen. A tungsten light may be used for illuminating the ground glass screen for making this final adjustment.

The 13 cm lens is then removed, the main diffraction lens, F_3, replaced and the diffraction pattern focused on the plane H without moving the micrograph holder G. It may be noted that a double hole specimen is also a good object for focusing the main diffraction lens onto the diffraction plane.

Important: When viewing the Airy disc directly the laser beam must be attenuated. This can be done using a suitable orientated Polaroid filter behind the microscope eyepiece because the laser beam is plane polarised.

REFERENCES

Bancroft, J. B., G. J. Hills and R. Markham (1967), A study of the self-assembly process in a small spherical virus, Virology *31*, 354.

Klug, A. and D. J. De Rosier (1966), Optical filtering of electron micrographs: reconstruction of one-sided images, Nature *212*, 29.

Lipson, S. G. and H. Lipson (1969), Optical physics (Cambridge University Press).

Chapter 5

The interpretation of optical diffraction patterns

5.1 Introduction

Optical diffraction patterns of electron micrographs are diffraction patterns obtained by using the photographic plate, or a copy of the plate, as a diffraction grating and as such can provide information only about the plate itself. The relevance of any such information to the structure of the specimen originally photographed in the electron microscope is limited to the extent that the actual structure is faithfully recorded in the picture.

In electron microscope studies of biological material it is generally necessary to use some type of non-specific staining procedure, such as negative staining, in order to achieve adequate specimen contrast. Negative staining involves the surrounding of a relatively transparent specimen by an electron opaque matrix. This results not in a true replica of the specimen but in a derivative structure, which is observed as a shadowgraph of those parts of the specimen not penetrated by the stain. This very fact greatly complicates the interpretation of images from negatively-stained specimens. In particular one may find that structural regulatities appear rotated in the image by 30° or 45° about a centre of symmetry (Bancroft et al. 1967; McLeod et al. 1963) in structures that have a fundamental hexagonal or tetragonal symmetry in the specimen plane. Furthermore since the object actually examined in the optical diffractometer is the electron micrograph plate, or a copy of it, the diffraction pattern obtained is characteristic of the plate rather than of the specimen photographed. Consequently optical diffraction patterns also detect such details as objective lens astigmatism, extent of defocussing, image movement during the exposure, non-linearity of the photographic process (see § 5.5.1), etc., as well as the peculiarities of the specimen preparation procedure. In addition, the area selecting aperture (see Appendix 1, E) or any mask placed around the area of the micrograph to be examined will influence the diffraction pattern.

5.2 The effects of the presence of non-periodic structures in the electron micrograph

Optical diffraction patterns have a further crucial deficiency when compared with electron or X-ray diffraction patterns. An optical diffraction pattern of some kind is produced by any two points on the plate under examination, and therefore the diffraction pattern records all the information present that can be recorded as amplitude variations on the plate. Put more simply, the optical diffraction procedure does not reject information from the non-periodic structures and so diffraction patterns are produced from both regular and non-regular arrays, and indeed by interference between rays originating from both types of structure. The result is that much of the diffraction pattern that appears to have originated from 'noise' in the micrograph may in fact contain information about the regular detail present. This is the main factor that limits the further extraction of information from the electron micrograph by procedures such as image reconstruction coupled with optical filtering. It is, of course, self evident that the greater the area occupied by regular structures and the greater the perfection of these structures, the sharper the associated diffraction spots will be, and the more meaningful the results that further processing may afford. Although it is not quite so obvious, it is also true that the artificial production of distinct arrays by photographic summation of similar objects can result in an improvement in the signal-to-noise ratio, always provided that the arrays are not identical. That is to say, it is important when summing regularities in structure to randomise the noise present. In general this is difficult to accomplish, although it might be done under stringent diffractometer control, but there are certain cases in which the symmetry elements of the object may be exploited for such a purpose. An example of this is an object with hexagonal symmetry in which repeated rotations of 60° will permit summation of regular detail while failing to sum noise. This is directly comparable to the rotation procedure of Markham et al. (1964) which can be applied to the electron micrograph before making the final diffraction pattern.

With the foregoing qualifications, the examination of optical diffraction

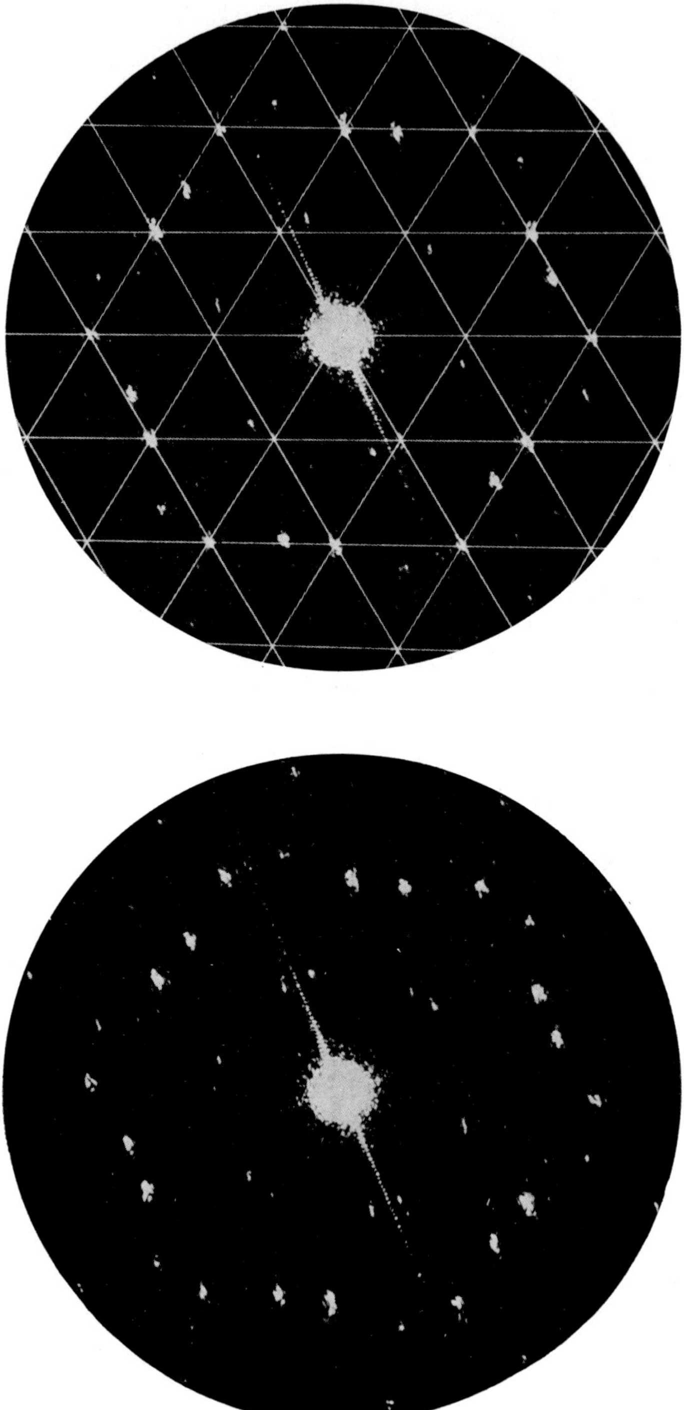

(a) (b)

Fig. 5.1 Identification of diffraction spots originating from two superimposed patterns. (a) The optical diffraction pattern of a tubular inclusion body from a turnip yellow mosaic virus infected plant. Notice that the sets of spots fall into two similar groups and are related to each other by a rotation of 14°. (b) Lines have been drawn through rows of spots corresponding to one of the sets mentioned above. Notice that all the spots in this group lie on the points of intersection of the lattice of equilateral triangles. The spots not so located lie on a similar lattice which is related to the former by a rotation of 14° in a clockwise direction. These sets originate from opposite sides of the tube. Reconstructed images using diffraction patterns isolated on the basis of the above analysis are shown in Fig. 5.2.

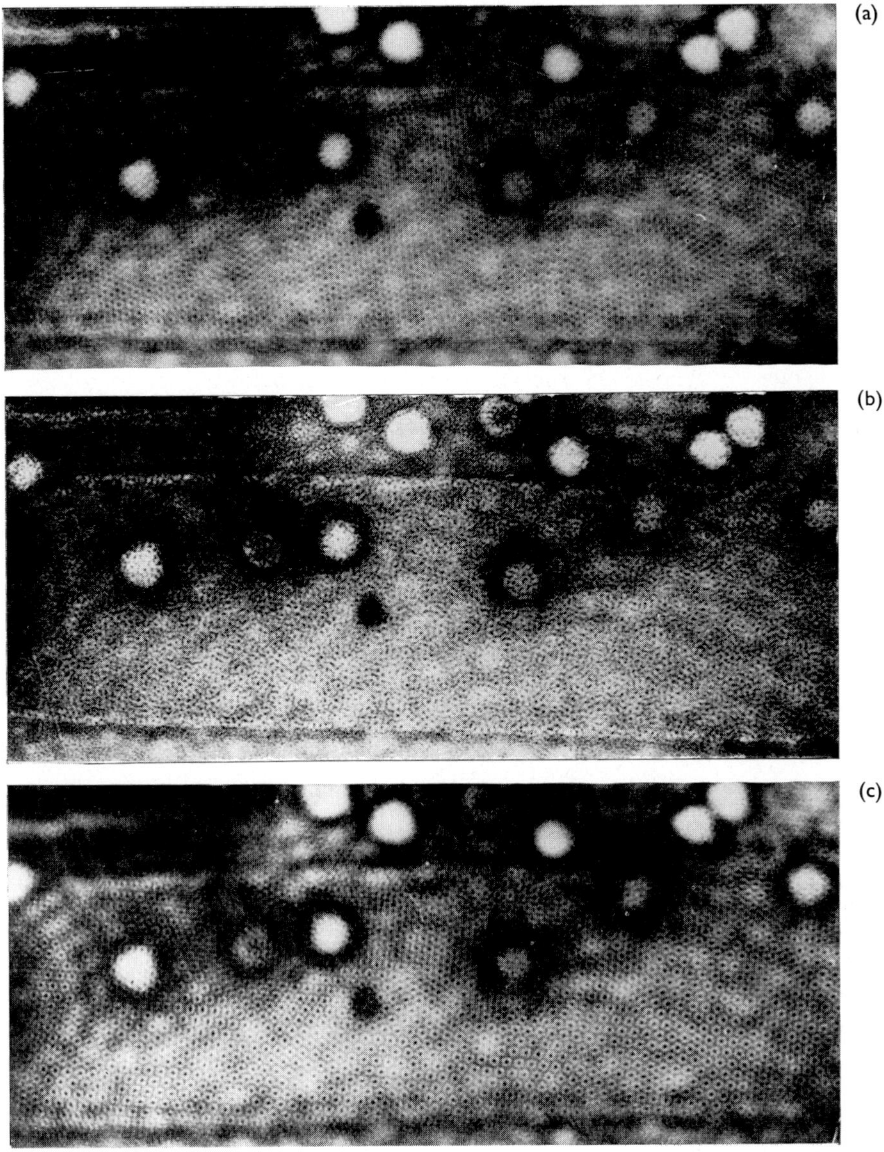

Fig. 5.2. Reconstructed images resulting from the selection of two different sets of spots in the diffraction pattern of an electron micrograph of a tubular inclusion body present in a turnip yellow mosaic virus preparation. (a) Reconstructed image using a selected set of spots from the diffraction pattern in Fig. 5.1, showing one side of the tubular inclusion. (b) Original electron micrograph showing Moiré features due to structural detail from both sides of the tube being imaged simultaneously. (c) Reconstructed image using the other set of spots from the diffraction pattern in Fig. 5.1. The pattern is related to that in (a) by a rotation through 14°.

patterns of electron micrographs is an exceedingly profitable enterprise. Regularities that cannot be distinguished by eye are often very easily seen in the diffraction pattern (see Fig. 4.8). This is particularly true when the object is double-sided and both sides are recorded superimposed in the microscope image (Figs. 5.1 and 5.2).

5.3 Analysis of periodic spacings and orientations in the electron micrograph

The determination of the periodic spacings and orientations in the electron micrograph by analysis of the optical diffraction pattern is precisely analogous to the determination of interplanar spacings and orientations in the specimen by analysis of the electron diffraction pattern (Beeston, §§ 4.1–4.2.1). The distance measured on the optical diffraction pattern between any particular first order diffraction spot (i.e. the first spot out from the centre along a row of spots) and the central spot of the pattern is inversely proportional to the periodic spacings between the regular features in the micrograph from which the diffraction spots originate. That is, the further the first order diffraction spot is from the centre, the finer is the spacing of the periodic detail from which the spot originates. In addition, the higher the orders of the spots that appear in the diffraction pattern, the finer is the detail in the corresponding spacing.

The equation relating the periodic spacing in the electron micrograph, d, and the distance between the associated first order spot in the optical diffraction pattern and the central spot, R, is, in exact analogy to electron diffraction,

$$Rd = \lambda L,$$

where λ is the wavelength of the laser light (632.8 nm for the helium neon laser), and L is the diffraction camera length. For an optical diffractometer with no lenses between the micrograph, G, and the diffraction plane, H (as in Appendix 1b and Fig. 1.3) the camera length is simply the physical distance, GH. However, if lenses are included between G and H, as in Appendix 1a, the camera length, L, is no longer the physical distance GH and must be determined using a test object with known spacings, d, by measuring the value of R from the optical diffraction pattern, and inserting the values in the above formula to obtain an accurate figure for L, in a very similar manner to determining the camera length for electron diffraction in the electron microscope (Beeston, § 5.1.3). A number of suitable test objects for calibration of the camera length of the optical diffractometer is listed in Table 6.1.

Once the camera length of the optical diffractometer has been calibrated, it remains fixed unless, of course, it becomes necessary to move any of the optical components or the micrograph holder, when it must be recalibrated for the new setting. If the components are frequently moved, as may be the case with simple systems, the camera length need not be determined each time, but a simple ratio technique, similar to that described in Beeston, § 4.2.4, may be used. A line grating test object, with known periodic spacing d_T, is used to form a diffraction pattern with the new setting of the optical system, and the distance between the diffraction spot and the central spot, R_T, is measured. The test object is now replaced by the micrograph to be analysed, taking care to ensure that it is placed in exactly the same plane as that occupied by the test object. The diffraction pattern is formed and the R value for the micrograph diffraction pattern, R_M, is measured. The periodic spacings, d_M, for the repeating structure of the micrograph is then given by

$$d_M = \frac{R_M}{R_T\, d_T}.$$

It must always be remembered, as already stressed, that the periodic spacings determined by optical diffraction analysis are those that are recorded on the electron micrograph, and any attempt to relate these spacings to the lattice spacings actually present in the microscope specimen are very dependent upon the accuracy with which the magnification of the electron microscope can be determined. This factor is discussed in detail in Chapter 6.

5.4 The use of optical diffractometers for the analysis of electron microscope image defects

As mentioned already, the optical diffraction pattern contains detail from the entire photographic plate or film and any image defects recorded in the electron micrograph, or faults in the plate itself, will influence the diffraction pattern. Consequently the presence of such image defects can be used to some advantage in order to determine the performance of the microscope. For these reasons some thought must be given to the quality of the image before it is subjected to analysis in the diffractometer.

Defects in the image, such as astigmatism, specimen drift and out-of-focus effects, have a characteristic effect on the 'noise' spectrum of the optical diffraction pattern, and it is only necessary to recognise these effects to diagnose imaging problems.

A micrograph having none of these defects, that is, taken exactly at focus and with no astigmatism or specimen drift present, will show a radially symmetric noise spectrum, slowly decreasing in intensity away from the central spot. This perfect noise spectrum will be altered in a characteristic way by the various image defects that may be present. If the micrograph is of an amorphous specimen area and contains no regular repeating detail, then only the noise spectrum will appear in the optical diffraction pattern. However, if the micrograph is of an area containing repeating structures then the optical diffraction will consist of a regular array of bright spots superimposed on the noise spectrum (Fig. 5.3).

5.4.1 ASTIGMATISM, SPECIMEN DRIFT AND EXTERNAL FIELDS

One of the most common image defects experienced by operators of the electron microscope is residual astigmatism of the objective lens. If suitable precautions are not taken to compensate the objective lens accurately before the final image is recorded, the microscope resolution will obviously be affected. Although most modern electron microscope objective lenses are fitted with efficient electrostatic or magnetic devices for the correction of astigmatism resulting from asymmetries in the lens magnetic field, the correction of residual astigmatism requires some skill and experience. The examination of a print, from what may have been considered to be a reasonable image recorded in the electron microscope, may not indicate any asymmetry in the original plate unless the plate was recorded at a sufficiently high magnification. However, when the photographic plate is examined in the optical diffractometer any residual asymmetry in the image will be indicated by the anisometric shape of the noise spectrum in the optical diffraction pattern. The diagrams in Fig. 5.3. illustrate the effects of the presence of typical residual astigmatism on the appearance of the noise spectrum from an electron microscope image recorded at high magnification. Measurement of the extremities of the noise spectrum shows that the image detail resolved is finer in one direction than the other.

Anisometric noise spectra also result from any small movement of the specimen or of the specimen stage due to thermal effects or to mechanical vibration of the microscope during the photographic exposure. In addition, any movement of the image due to changing external magentic fields will cause anisometric noise spectra to appear.

5.4.2 OUT-OF-FOCUS EFFECTS

It is common practice to defocus the electron microscope objective lens slightly, generally by underfocussing, to achieve phase-contrast conditions

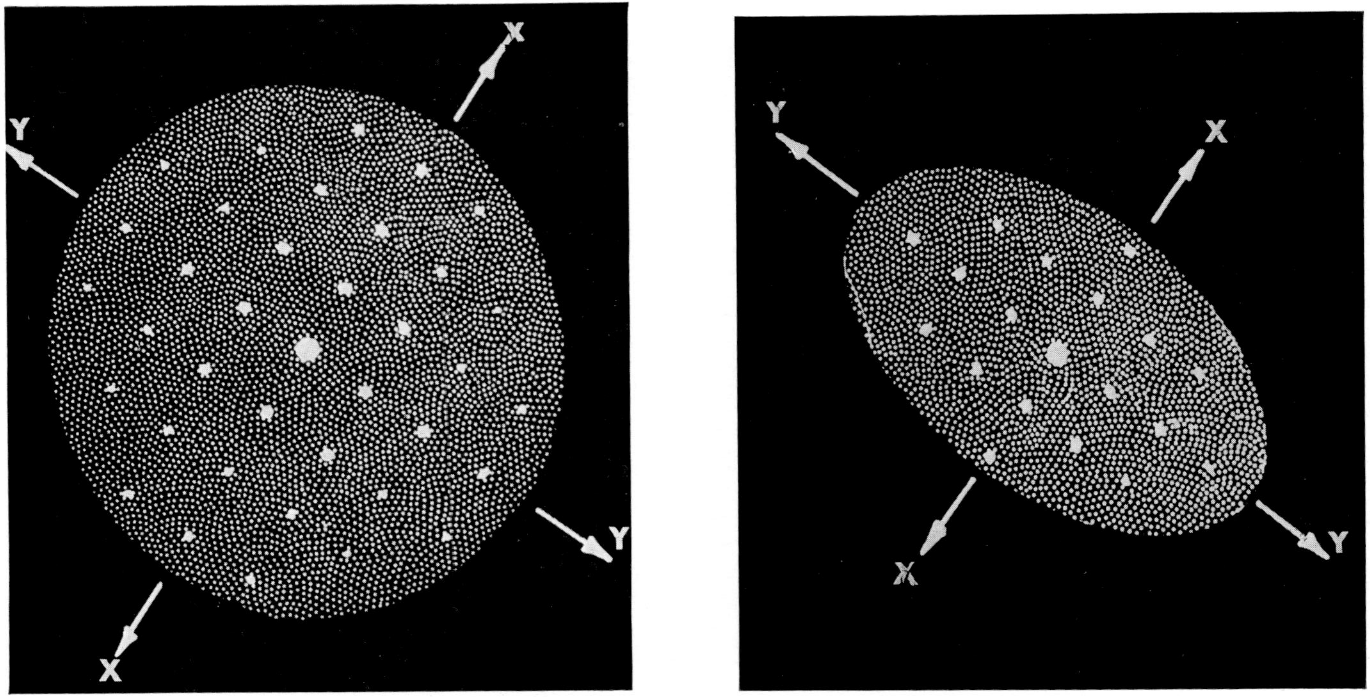

(a) (b)

Fig. 5.3 The diagrams represent optical diffraction patterns obtained from electron micrographs possessing regular repeating features. The larger spots indicate a typical pattern from a regular lattice arrangement, whereas the smaller spots are produced from irregularities or noise simultaneously recorded in the original electron microscope image. In the case of a corrected lens free from astigmatism the noise spectrum will appear as indicated in (a), and the resolution will be the same in both directions X and Y. Electron micrographs with astigmatism will produce a noise pattern as (b), with an elongated optical diffraction pattern in the direction Y and corresponding reduction of resolution in direction X.

which enhance structural details at particular spacings in the final image and increase contrast. The optimal conditions for using phase-contrast effects in this way are limited, and if the extent of defocusing approaches 500 to 1000 nm artefacts may be introduced in the final image, such as fringe effects or contrast reversal. The particular spacings in the image that are enhanced by certain amounts of defocusing can be clearly observed by forming an optical diffraction pattern of the recorded micrograph (Thon 1967; Ericson and Klug 1971). Since the distance out from the centre of the optical diffraction pattern at which the bright feature appears is a measure of the interparticle or interplanar spacings from which that feature originates in the specimen, then, if certain spacings are enhanced by defocusing, these spacings will appear as bright rings or bands in the noise spectrum in the optical diffraction pattern (Fig. 1.1). By measuring the radius of one of these rings, R, it is a simple matter to determine the spacing that is enhanced in the image, d, by using the formula

$$R\,d = \lambda\,L$$

as indicated in § 5.3.

If more than one ring appears in the optical diffraction pattern, then a number of different spacings are enhanced in the recorded image, and the size of these spacings can be determined by measuring each radius in turn and applying the formula above. If the ring appears more as a broad band than as a sharp ring then a range of spacings is enhanced in the image; the range can be determined by measuring the inner and outer radii of the band.

5.4.3 DETERMINATION OF INSTRUMENTAL RESOLUTION

The ultimate test and measurement of the resolution of the electron microscope comes from analysing a micrograph of a test object in the optical diffractometer. For an amorphous test object, such as a fine carbon film coated with evaporated metal particles, the resolution is taken as a point-to-point measurement, when the objective lens is at focus. An optical diffraction pattern of a micrograph of such a test object will show a bright noise spectrum symmetrically distributed about the central spot, as in Fig. 5.3a. Measurement of the radius of the extremity of the bright noise disc, R, will give a measure of the resolution of the microscope, that is the smallest interparticle spacing, d, that will be resolved, at the time the micrograph was recorded, by using the standard formula

$$R\,d = \lambda\,L$$

if the camera length, L, is known with reasonable accuracy.

A further critical evaluation of the resolving power of the electron microscope may be made from a measurement of the linear resolution using one of the crystal lattice structures used for calibration purposes (see Table 6.1).

At high resolution the contributions of spherical aberration, phase contrast and amplitude contrast to image formation in the electron microscope are relatively complex and the reader is referred to the articles of Thon (1965, 1967), Heidenreich (1967) and Ericson and Klug (1971).

5.5 Photographic emulsions and plate defects

It is clear that apart from the random noise present in the electron microscope image, the graininess of the photographic plate exposed to the electron beam must be considered as an additional noise factor that will limit the resolution of the recorded micrograph.

The choice of suitable photographic material for the electron microscope has led to a number of controversial discussions (see Valentine 1965, 1966; Akashi et al. 1958; Highton 1965), about the grain sizes in different emulsions exposed to electrons within the range of 20 to 100 kV. For all emulsions any defects or irregularities on the plate surface will be seen when the plate is placed in the diffractometer. Changes in glass plate thickness or unparallel surfaces can act as an additional lens in the diffractometer system and effectively change the focal position of the optical diffraction pattern. It has been our experience that high quality contact prints made directly onto film are the most satisfactory for examination and measurement in the diffractometer, provided that suitable precautions are taken to keep the film reasonably flat in the mount. In general it is not necessary to mount films between optical flats for examination, though this may be necessary in the case of some plates.

5.5.1 THE EFFECTS OF THE PHOTOGRAPHIC PROCESS

At the moment it is not possible to examine objects directly in the electron microscope. Before any image can be formed the electron beam must be subjected to a detection procedure. There are many possibilities for such detection, but in electron microscopes, other than some scanning microscopes, only photography is used as a routine. Photographic detection, like most other practicable methods measures intensity only. That is to say it is insensitive to phase differences and responds to the square of the amplitude of the electron beam, which is always, of necessity, a positive function. In order to produce a true representation of the structure of the object, it is essential that the relationship between the logarithm of the electron beam intensity and the time of exposure should be linear with respect to the photographic density in the developed plate. In other words the transfer function should be linear. This is quite evidently not so. The effective visible contrast on the photographic plate is dependent on the exposure time and the contrast is far from uniform in any given image. Valentine (1965) has suggested that the sensitivity of photographic plates towards electrons is linear. Though this may not necessarily be true, it is certainly true that the desirable logarithmic relationship between exposure and final density does not hold for electrons. What effects will this have on the interpretation of electron micrographs? If one considers a grating having a sinusoidal distribution of matter, or a dense linear grating at the limit of resolution, it is evident that the resulting sinusoidal modulation of the electron beam intensity will be distorted on detection. Taking the theoretical curve of Valentine (his fig. 1) for a one-hit process, it would appear that the distortion would be mainly second harmonic at low final densities. What effects would this have on the optical diffraction pattern? The simple answer is that since the diffractometer is a very sensitive harmonic analyser one must expect spurious harmonic detail. In other words the microscope will probably appear to be at least twice as good as it really is.

Since it is often inconvenient to examine the original plate in the diffractometer, ordinary photography is frequently used to make copy positives or negatives from it. Although the relationship between density and the logarithm of the photographic exposure is more nearly linear in this case, all photographic materials also have a 'toe' region that may introduce further distortion.

As has been mentioned, non-linear transfer of information to the photographic material has the effect of producing spurious detail. How can this be recognised? The answer is that in general it cannot be except in certain specific cases. Even harmonic distortion, and it is almost entirely second harmonic distortion that might be encountered, can be suspected when weak second order spots are present in optical diffraction patterns. These spots will lie on a line joining the centre of the pattern and a strong spot, and be twice as far from the centre as the strong spots (see, for example, Fig. 6.3). Note that the second order spots due to this cause are *always* weaker. So spots that have no stronger spot half way between them and the centre of the diffraction pattern in a straight line can be regarded as genuine. By using such spots as landmarks, it is simple to decide at what radius detail is likely to become dubious.

In conclusion it is probably sensible to comment that it is only when one is stretching technique to the utmost that photographic distortion will prove to be a source of error. On the other hand if one should be examining very high contrast prints or reconstructed or integrated images, one must expect to find quite serious harmonic distortion.

5.6 Periodic spacing requirements of the electron micrograph

The final requirement of the micrograph to be analysed in the optical diffractometer is that the periodic spacings or interparticle distances in the micrograph be within a size range that will give spots or rings that lie within the bounds of the viewing screen or photographic plate of the optical diffractometer, and are not so close to the central spot as to be obscured by the scatter caused by the intensity of the bright central spot.

If the optical diffractometer has an infinitely variable camera length, as may be the case in the system shown in Appendix 1 if the ground glass screen or plate holder (at H) is free standing and a long bench or corridor is available, then there is no limitation on the spacings that can be accommodated. However, the full optical systems generally have a limited range of camera tengths, L, and a limited size of viewing screen. More important is the need to expand the diffraction pattern to a reasonable size for the ease of making the filtering mask, if optical filtering is to be used. This places a lower limit on the value of R of a few millimetres.

The upper limit of R depends upon the size of the camera format, but will be about 60 mm if a Polaroid Land camera is used, and much less if 35 mm recording is employed.

Knowing the limits of the value of R that can be accommodated, and the limits of the range of camera lengths, L, that are available, it is a simple matter to determine the range of periodic and interparticle distances, d, that will give a useful diffraction pattern, by using the standard formula

$$Rd = \lambda L$$

where $\lambda = 632.8$ nm for helium neon laser light. To take the specific example of the Talbot diffractometer (§ 7.4.2), in which the camera length, L, is fixed at 880 mm, and the maximum value of R that can be accommodated on the filtering mask is about 9 mm, it is clear that the size of the spacings, d, in the micrograph must not be less than

$$d = \frac{\lambda L}{R} = \frac{632.8 \times 880}{9} = 0.06 \text{ mm (approx.)}$$

if the diffraction spots are to appear within the bounds of the filtering mask.

In addition, the spacings must not be too much greater than this, or the spots will be very close together in the diffraction pattern. It is clear that any detail in the microscope image which is to be analysed in the Talbot diffractometer must be recorded at such a magnification that the spacings on the plate are less than, and fairly close to, 0.07 mm. If the plate has not been recorded to meet this requirement, then a copy should be made onto film, such that the final spacings do satisfy this limitation.

REFERENCES

Akashi, K., T. Masuda, H. Tochigi, K. Yamanouchi and E. Iguchi (1958), The effect of grain of photographic emulsion on the resolution of an electron microscope, in: Fourth Internal Conference on Electron Microscopy, Berlin (Springer-Verlag), p. 131.

Bancroft, J. B., G. J. Hills and R. Markham (1967), A study of the self-assembly process in a small spherical virus, Virology *31*, 354.

Ericson, H. P. and A. Klug (1971), Measurement and compensation of defocusing and aberrations by Fourier processing of electron micrographs, Phil. Trans. Roy. Soc. Lond. *B261*, 105.

Heidenreich, R. D. (1967), Electron phase contrast images of molecular detail, Siemens Rev. *34*, 4.

Highton, P. (1965), The measurement of dimensions in the image plane of the electron microscope, in: Quantitative electron microscopy, G. F. Bahr and E. H. Zeitler, eds. (Williams and Wilkins, Baltimore), p. 171.

Macleod, R., G. J. Hills and R. Markham (1963), Formation of true three-dimensional crystals of the tobacco mosaic virus protein, Nature *200*, 932.

Markham, R., J. H. Hitchborn, G. J. Hills and S. Frey (1964), The anatomy of tobacco mosaic virus, Virology *22*, 342.

Thon, F. (1965), Elektronenmikroskopische Untersuchungen an dünnen Kohlefolien, Z. Naturforsch. *20A*, 154.

Thon, F. (1967), On the defocusing dependence of phase contrast in electron microscope image, Siemens Rev. *4*, 13.

Valentine, R. C. (1965), Characteristics of emulsions for electron microscopy, in: Quantitative electron microscopy, G. F. Bahr and E. H. Zeitler, eds. (Williams and Wilkins, Baltimore), p. 596.

Valentine, R. C. (1966), The response of photographic emulsions to electrons, in: Advances in optical and electron microscopy, vol. 1, R. Barer and V. E. Cosslett, eds. (Academic Press, London), p. 180.

Chapter 6

Calibration requirements and test objects

6.1 Causes of variation in magnification of the electron microscope

One of the important applications of the optical diffractometer to electron micrographs is the determination of spacings and the dimensions of repeating features appearing in the optical diffraction patterns. Accurate measurement, however, will depend on the correctness of the instrumental magnification value at which the original micrograph was recorded. Precise calibration of the optical diffractometer is not difficult and can be performed by the use of several test objects which will be discussed in more detail later, but calibration of the electron microscope is less easy and requires further comment here.

For the majority of the biological problems studied by electron microscopy the accuracy of the magnification values is not critical and in many publications there is little reference to the methods used to determine the variations in magnification with time of a given instrument. In practice the manafacturers give a magnification tolerance of about $\pm$ 5%, but when commercial instruments are subjected to critical tests with suitable magnification standards, this tolerance may be found to be much higher.

The reasons for variations in the magnification with time of electron microscopes have been discussed by Elbers and Peiters (1964), who showed that variations in the accelerating voltage, and lens current settings for different positions of the specimen when in focus, together with lens hysteresis, may have serious effects on magnification accuracy. Other causes for variations of magnification associated with certain instruments were described by Cox and Horne (1968), who drew attention to the chromatic changes in magnification that may occur across the recorded image at various settings of the intermediate and projector lenses when these are used at the extreme ends of the magnification range.

6.2 Calibration at low magnification

Various methods have been described in the literature for the calibration of the lower range of magnifications (5000 × to 15,000 ×) (see Hall 1953). One of the most popular internal standards for the determination of instrumental magnification is the grating replica* which has been discussed in detail by

TABLE 6.1

Magnification and calibration data for electron microscopy

Conversion table

Units	*Inch*	*cm*	*mm*	*µm* *micrometre*	*nm* *Nanometre millimicro-metre*	*Å* *Angstrom*
Inch	1	2.54	25.4	25,400	25,400,000	254,000,000
cm	0.393	1	10	10^4	10^7	10^8
mm	0.0393	10^{-1}	1	10^3	10^6	10^7
µ	0.000,039,3	10^{-4}	10^{-3}	1	10^3	10^4
mµ nm	0.000,000,039,3	10^{-7}	10^{-6}	10^{-3}	1	10
Å	0.000,000,003,93	10^{-8}	10^{-7}	10^{-4}	10^{-1}	1

Lattice spacings for internal standards			*Zone axis*
Pt. Phthalocyanine	12 Å	1.2 nm	[201]
Cu. Phthalocyanine	14.37 Å	1.437 nm	[111]
Na. Faujasite	11.5 Å	1.11 nm	
Tremolite	3.2 Å	0.32 nm	
Au. Lattice	2.35 Å	0.235 nm	[111]
Au. Lattice	2.04 Å	0.204 nm	[200]
Catalase A Spacing	62.1 Å	6.21 nm	
Catalase B Spacing	81.2 Å	8.12 nm	

Grating replicas

A	1500	Lines/inch	55	Lines/mm
B	28,800	Lines/inch	1134	Lines/mm
C	54,860	Lines/inch	2160	Lines/mm

* See List of suppliers.

Bahr and Zeitler (1965). An optical cross-grating is made by accurately ruling two gratings of 2160 lines/mm (see Table 6.1) at right angles to each other. A metal replica is made of this grating, and then a plastic replica is produced from the metal replica. The plastic replica is shadowed in the two directions at right angles to the grating rulings. Shadowing is necessary to enhance the edges of the grating lines and allows greater accuracy of measurement.

Grating replicas may suffer to some extent from distortion in the electron beam and the image may be distorted at low magnifications in certain projector lens systems. In microscopes not fitted with adequate anti-contamination devices it is possible to obscure the edges of the gratings by heavy carbon deposits. The problem of contamination of the specimen also occurs with any object being used as an internal standard for the purposes of magnification calibration.

When micrographs are recorded at 20,000 × and above it is obvious that the number of spacings of a grating replica within a given area on the viewing screen in the microscope will fall off rapidly and consequently a large number of micrographs must be recorded and measured in order to obtain a good value for statistical analysis, and it is preferable to use another method of calibration.

6.3 The use of catalase enzyme crystals for calibration

In many biological structures suitable for analysis in the optical diffractometer spacings of about 2.0 to 10.0 nm are frequently encountered. Other specimens, however, contain repeating features of much larger spacings and for calibration purposes an internal standard suitable for a wide range of magnifications is desirable.

One of the most successful test objects for the calibration of a wide range of magnifications is the catalase enzyme crystal. Microcrystals of catalase prepared by the negative-staining method were shown by Horne and Greville (1964) and Horne (1965) to exhibit regular lattice spacings of about 6.7 nm at right angles to the long axis of the crystal and about 8.7 nm parallel to the axis (Table 6.2). Crystalline preparations of catalase are commercially available (see List of Suppliers). In a more recent detailed analysis of catalase as an internal magnification standard it was shown by Wrigley (1968) that these microcrystals provide stable and consistent objects which are relatively free from irradiation damage when illuminated under normal double condenser and electron beam conditions. Values for the spacings in catalase

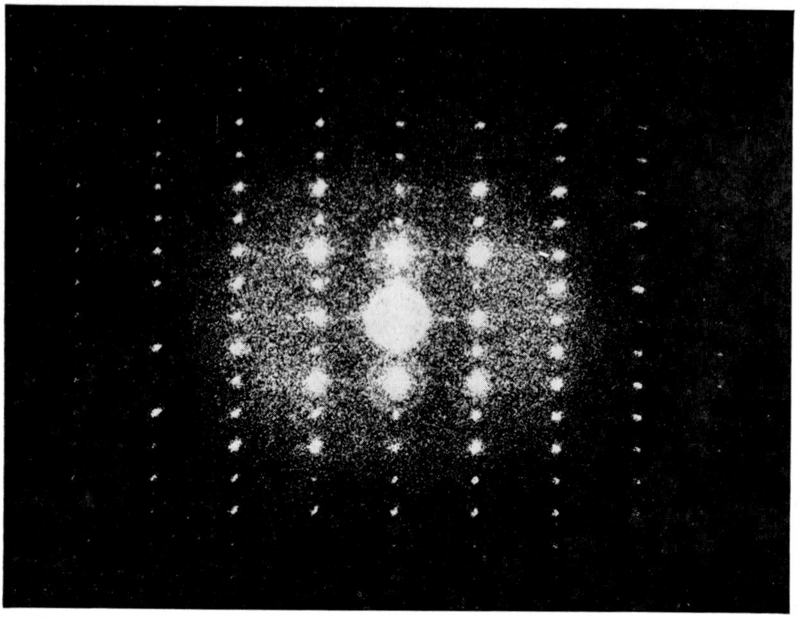

(a)

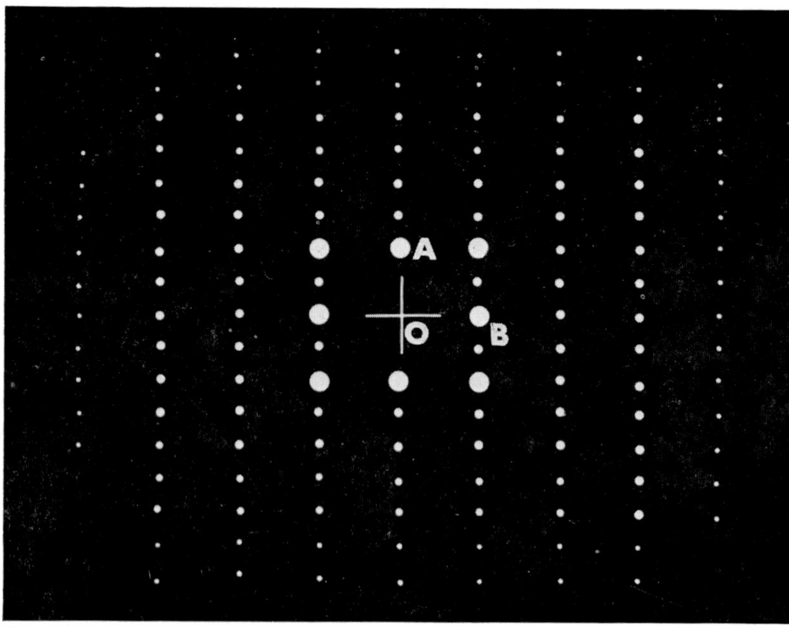

(b)

crystals reported by several workers are summarised in Table 6.2. The optical diffraction pattern from a region of a catalase crystal is shown in Fig. 6.1a. It can be seen that there are several orders of diffraction spots; the spacings are indicated in the diagram of the diffraction pattern (Fig. 6.1b). These dimensions are similar in range to many of those observed in biological objects studied in the electron microscope.

6.4 Calibration at high magnification

It has been established that the most accurate method for calibration of the electron microscope when working at high instrumental magnifications (80,000 × to 200,000 ×, or more) is to measure the lattice spacings of certain inorganic microcrystals which have already been determined by X-ray diffraction analysis and which can be visualized directly in the transmission electron microscope. Various microcrystals and thin metal foils have spacings of the values shown in Table 6.1 and can be viewed in the electron microscope. The advantages of these specimens as internal standards are twofold: firstly, there are a relatively large number of measurements available from a single micrograph at high magnification settings of the microscope, and secondly, the crystals can be mixed with many biological specimens in suspension, or added to the support film. They also have certain advantages when used in conjunction with the optical diffractometer when the diffractometer is set to a given range of magnification or camera length. The size of the diffraction pattern will depend on the lattice spacings in the specimen, the magnification at which the plate is recorded and the optical arrangement of the diffractometer bench. For these reasons some consideration has to be given to the basic magnification at which the electron micrograph is recorded and to the spacings to be measured with the optical diffractometer (§ 5.6). A typical diffraction pattern from a micrograph of a platinum phthalocyanine crystal (see Menter 1956) is shown in Fig. 6.2.

One of the disadvantages of using microcrystals of Cu or Pt phthalocyanine is the problem of irradiation damage to the crystals due to the electron beam. Obviously, in order to focus the image accurately at high magnifica-

Fig. 6.1. (a) An optical diffraction pattern from an electron micrograph of a negatively stained catalase crystal. The spacings and angles of the catalase crystal lattice are well defined in the diffraction pattern. In the print shown the noise spectrum is under exposed and would normally extend beyond the 4th order spots. (b) A tracing from the diffraction pattern in (a). The basic spacings are OA = 8.15 nm and OB = 6.21 nm. It can be seen that repetitive detail in the micrograph is limited to a resolution of about 1.2 nm.

TABLE 6.2

Measurements of the lattice spacings of catalase enzyme crystals. (Reproduced from data supplied by Dr. N. C. Wrigley.)

Reference	*Value of* $\frac{1}{2} C_0$*	*Std. dev.* ± *(nm)*	*No of crystals measured*	*Spacings measured per crystal*	*Crystal preparation*	*Technique*
Hall 1950	7.95	?	?	up to 200	Shadowed replicas	Direct image, calibration unknown
Labaw 1967	9.15	0.15	50	up to 100	Shadowed replicas	Direct image, precalibration with Olive T crystals
Luftig 1967	8.8	0.3	66	Over 80	Unfixed, KPT and UA negative stains.	Direct image, precalibration with Cu phthalocyanin crystals
Kiselev et al. 1967	8.7				Unfixed, KPT (various pH) and AM negative stains	Direct image
Ferrier and Murray 1966	8.15	0.2	70	Over 1000	Unfixed, AM negative stain.	Electron diffraction, internal calibration with Au and Al films
Valentine 1967 (unpublished)	8.25	0.2	27	Over 1000	Same material: Glutaraldehyde-fixed, SST negative stain.	Electron diffraction, internal calibration with Au film
Wrigley 1968	8.75	0.1	24	Over 1000		Electron diffraction, internal calibration with Au film
Cox and Horne 1968	8.44	0.2	50	Over 80	Negative staining	Direct image, and microdensitometer

Abbreviations:
KPT: Potassium phosphotungstate, UA: Uranyl acetate, AM: Ammonium molybdate,
SST: Sodium silicotungstate.
* $\frac{1}{2}C_0$ = Simple mean for crystal lattice spacing (see Wrigley 1968)

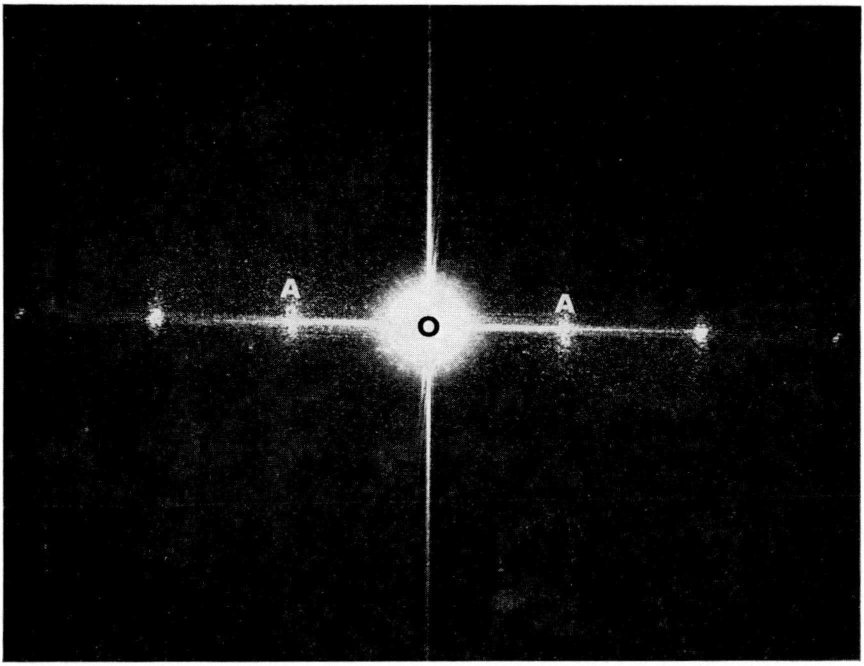

Fig. 6.2. Optical diffraction pattern from an electron micrograph of a platinum phthalocyanine crystal. The spots indicated at A show the 1.2 nm lattice spacing. (Kindly supplied by G. J. Hills, John Innes Institute.)

tion adequate illumination is required with an increase in the risk of specimen damage, and in practice the operator may have to guess the correct focus position and then take a through focal series of micrographs from a given crystal illuminated at relatively low beam intensity.

In the very high magnification ranges of maximum resolution instruments (200,000 × to 500,000 ×) it also possible to use the lattice spacings in thin metal foils, such as gold, for calibration (see Table 6.1). The methods of preparation of suitable foils can be found in the publication by Hirsch et al. (1965). Some of these foils are supplied by the manafacturers of electron microscopes (see List of suppliers). The optical diffraction pattern shown in Fig. 6.3 illustrates the spacings resolved in two directions in an electron micrograph of the gold foil specimen shown in Fig. 8.6.

It should be mentioned at this point that, in addition to magnification calibration, the specimens described above for high resolution also provide a good test for the performance of the electron microscope.

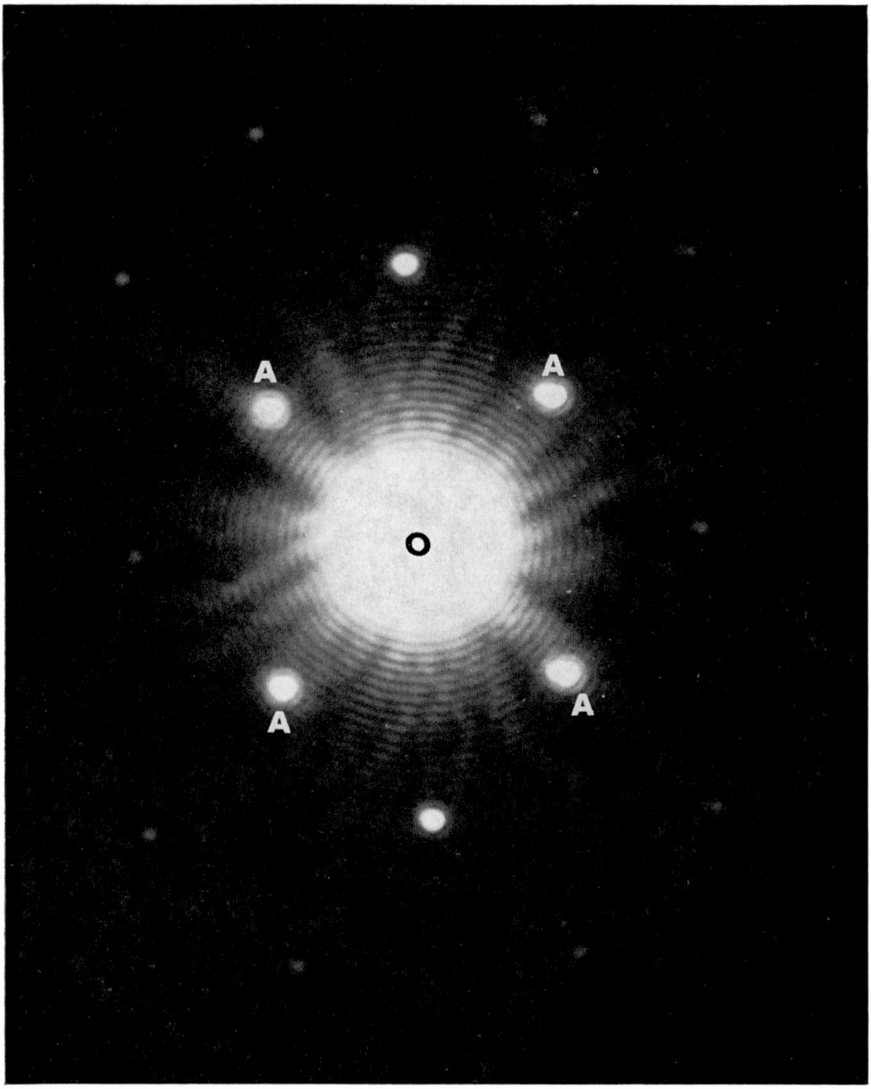

Fig. 6.3. Optical diffraction pattern from an area of the gold lattice structure shown in Fig. 8.6 (inset). The spots indicated at A show the lattice spacing of 0.204 nm.

6.5 Calibration of the optical diffractometer

Various optical gratings with accurate spacings are available for the calibration of the optical system of the diffractometer, and some of these are listed, together with their spacings, in Table 6.1. Similar gratings are also suitable for critical focusing of the lens systems of the diffractometer, as discussed in § 4.5.1. Copper meshes can be used for the same purpose (see List of suppliers).

Micrographs of catalase crystals are also most suitable for the calibration of the optical diffractometer, assuming that accurate determination of the electron microscope magnification has been carried out beforehand.

REFERENCES

Bahr, G. F. and E. Zeitler (1965), The determination of magnification in the electron microscope, in: Quantitative electron microscopy, G. F. Bahr and E. Zeitler, eds (Williams and Wilkins, Baltimore), p. 142.

Cox, R. W. and R. W. Horne (1968), Accurate calibration of the magnification of the A.E.I. EM 6B/2 electron microscope using catalase crystals. Fourth European Regional Conference on Electron Microscopy. Vol. 1 (Tipografia Poliglotta Vaticana, Rome), p. 579.

Elbers, P. F. and J. Pieters (1964), Accurate determination of magnification in the electron microscope, J. Ultrastruct. Res. *11*, 25.

Ferrier, R. P. and R. T. Murray (1966), Low-angle electron diffraction, J. Roy. Microsc. Soc. *85*, 323.

Hall, C. E. (1950), Electron microscopy of crystalline catalase, J. Biochem. *185*, 794.

Hall, C. E. (1953), Introduction to electron microscopy (McGraw-Hill, London).

Hirsch, P. B., A. Howie, R. B. Nicholson, D. W. Pashley and M. J. Whelan (1965), Electron micrographs of thin crystals (Butterworths, London).

Horne, R. W. (1965), The application of negative staining methods to quantitative electron microscopy, in: Quantitative electron microscopy, G. F. Bahr and E. Zeitler, eds. (Williams and Wilkins, Baltimore), p. 316.

Horne, R. W. and G. D. Greville (1964), Electron microscope studies of the structure of enzyme molecules. Third European Regional Conference on Electron Microscopy, Prague, Vol. B, p. 53.

Kiselev, N. A., C. L. Shpitzberg and B. K. Vainshtein (1967), Crystallization of catalase in the form of tubes with monomolecular walls, J. Mol. Biol. *25*, 433.

Labaw, L. W. (1967), Ox-liver catalase enzyme crystal structure by electron microscopy, J. Ultrastr. Res. *17*, 327.

Luftig, R. (1967), An accurate measurement of the catalase crystal period and its use as an internal marker for electron microscopy, J. Ultrastruct. Res. *20*, 91.

Menter, J. W. (1956), The direct study by electron microscopy of crystal lattices and their imperfections, Proc. Roy. Soc. Lond. *A236*, 119.

Wrigley, N. G. (1968), The lattice spacings of crystalline catalase as an internal standard of length in electron microscopy, J. Ultrastruct. Res. *24*, 454.

Chapter 7

Optical diffraction patterns from the conventional light microscope: convolution camera: commercial optical diffractometers

7.1 Conversion of the light microscope for optical diffraction

Although the technique of applying optical diffraction to the analysis of electron micrographs is more versatile with optical benches, it is possible to use the light microscope for the production of an optical diffraction pattern, as described by Gall (1967) who demonstrated that the conventional lens system in the compound microscope could be modified simply by the insertion of a small electron microscope aperture above the field diaphragm of the condenser as shown in Fig. 7.1. The aperture used was a standard 100 μm diameter platinum aperture disc. In addition, a sliding plate, which carries the micrograph and which can be moved across the optical axis, has to be inserted in the body tube of the microscope in the position labelled 'diffraction mask' in Fig. 7.1.

The production of a diffraction pattern in the light microscope from a suitable electron micrograph is a relatively simple matter, but some attention needs to be paid to the micrograph to be inserted in the system for the reasons given below. Obviously the standard plate on which the original micrograph is recorded is unsuitable and the selected area of interest must be copied photographically. The method recommended by Gall (1967) was to enlarge the electron micrograph onto photographic paper to allow careful

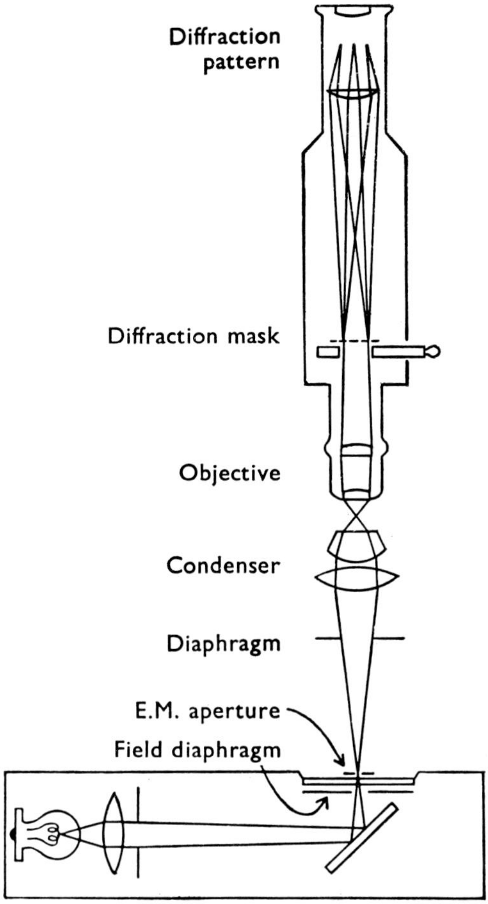

Fig. 7.1. The diagram shows a light microscope modified to observe the optical diffraction pattern from an electron micrograph. A small copy of the original micrograph or 'mask' is placed in the position shown and its diffraction pattern viewed through a conventional eyepiece. A standard electron microscope objective aperture is placed between the illumination field diaphragm and condenser substage diaphragm to provide suitable illuminating conditions. (From J. G. Gall 1967.)

examination of the detail and its orientation. Four strips of white paper or tape were then placed accurately at 90° to each other to outline the area of interest; two of these strips were placed parallel to the axis of the repeating morphological features to be analysed. The area outlined by the white strips of paper or tape was rephotographed onto 35 mm film to give a rectangular negative with a side of not more than 5 mm. This dimension of 5 mm was the maximum micrograph size for the particular type of optical system and microscope employed. The repetitive features in the reduced micrograph had spacings within the range of 25 to 500 μm. Various high contrast photographic emulsions are available and Kodak High Contrast Copy Film (see List of suppliers) was recommended by Gall (1967) as being most suitable.

The micrograph was cut to size, taped onto the sliding plate in the body tube of the microscope and the diffraction pattern viewed in a standard eyepiece (Fig. 7.1). It should be remembered that the size and spacing of the spots observed in the eyepiece will depend on the magnification of the microscope objective. For these reasons it was suggested that the use of a 10 × objective provided a good compromise between spot size and light intensity.

Alignment of the optical system and illumination was not found to be critical and suitable diffraction patterns were observed by the use of a standard 6 volt tungsten filament bulb with the field diaphragm set so as to illuminate an area just smaller than the full objective aperture. It is a simple matter to insert a standard 100 mesh electron microscope grid on the sliding plate as a test object and to observe the diffraction pattern in the eyepiece as described above. For the reasons mentioned in Chapter 2.2 some consideration must be given to the use of non-coherent and non-chromatic illumination sources, and the effects they will have on the diffraction pattern. It is possible to improve the coherence of the system by providing illumination from a mercury arc source.

The results published by Gall (1967) were similar to those described by Klug and Berger (1964) and Berger et al. (1966), who employed a more sophisticated optical bench, but there are a number of practical problems in the production of suitable small copy micrographs when large numbers of electron micrographs have to be subjected to routine examination using a light microscope arrangement. In addition, there are other limitations arising from the small size of the diffraction pattern when attempting to reconstruct the image. Nevertheless, this adaption of the light microscope can be of value in a small laboratory where the production of optical diffraction patterns from a few electron micrographs may be of interest.

7.2 Modification of a commercial optical bench for optical diffraction and filtering

It is possible to carry out relatively simple modifications to certain commercial optical benches for the purposes of optical diffraction and image filtering

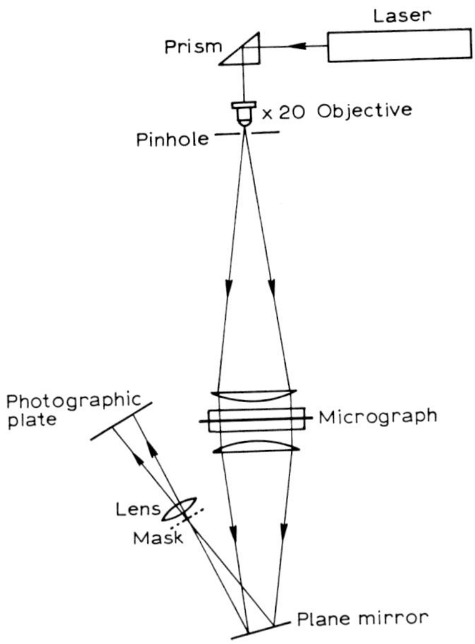

Fig. 7.2. The diagram shows the Pullin optical bench modified as described by Fraser and Millward (1970) for optical diffraction and filtering of electron micrographs.

as described by Fraser and Millward (1970). These authors claimed that image averaging techniques could be performed using this optical diffractometer by inserting layer line filtering masks in the optical system and then reconstructing the image, thus improving the signal-to-noise ratio by the removal of spots between the layer lines. The optical diffractometer used in their experiements was a modified Pullin optical bench and the experimental arrangement is shown in Fig. 7.2. Selected electron micrographs possessing repetitive detail were copied at suitable magnifications onto plates and secured between two optical flats using cedar wood oil for optical contact.

7.3 Convolution camera applied to electron micrographs

The optical process referred to as 'convoluting' is of some importance in the formation of diffraction patterns and can be demonstrated mathematically and experimentally. Convolution can be applied to objects possessing arrays, regular or otherwise, and for a detailed discussion of convolution theory the reader is referred to the publication on optical physics by Lipson and Lipson (1969). For the purposes of analysing electron micrographs it is possible to modify an optical bench and construct a convolution camera of the type illustrated in Fig. 7.3. This apparatus was described by Hosemann and Bagchi (1962) and has been used to reveal certain periodicities in electron micrographs which could not be detected by optical diffraction (Elliott and Lowy 1970; Squire 1971). The formation of the image on the photographic film or screen shown in Fig. 7.3 represents the convolution square of the original intensity distribution within the electron micrograph and can be viewed directly using the convolution camera. This process does not require the use of diffraction methods to detect periodicities in electron micrographs.

The arrangement shown in Fig. 7.3 is essentially a fixed focus camera with the viewing screen or film located in the posterior focal plane of the lens. In the case of the convolution camera described by Elliott et al. (1968), the lens was an f/2.5 anastigmat photographic lens. Long focal length telescope lenses (100 cm focal length) can be used in constructing a convolution camera, but as Elliott, Lowy and Squire point out, diffraction effects from such lenses can introduce blurring of the convolution square image. For this reason a shorter focal length lens is recommended.

Two identical copies of the electron micrographs or 'masks' to be examined are made at a suitable magnification. One is placed in front of the lens as shown in Fig. 7.3 and the other placed at a distance from the first equal to the focal length of the lens. These 'masks' must be carefully aligned with respect to each other. A ground glass diffuser is placed between the light source and the second electron micrograph. The light rays parallel to the axis will pass through each micrograph and come to a focus on the film or screen at N to give an axial spot of intensity proportional to the number of regular repeating features and intensity of interparticle vectors of zero magnitude. Other rays will pass through features on one micrograph and then through the other to produce an image at some point at a distance away from the zero order which will be equal to the interparticle spacing or basic periodicity with an intensity equal to $N-1$.

7.4 Commercial optical diffractometers

There are many optical bench systems available from commercial sources which can easily be modified for the purposes of analysing electron micrographs and the basic apparatus already described in earlier sections of this volume is directly applicable for obtaining diffraction patterns. It is not necessary, nor our intention, to include a comprehensive list of suppliers of

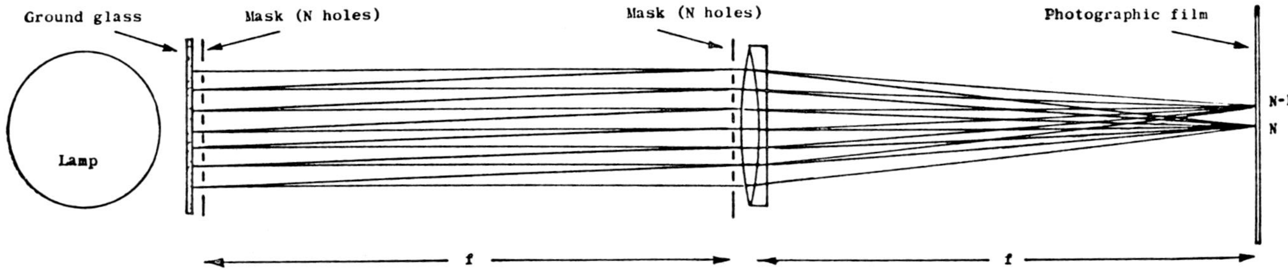

Fig. 7.3. The diagram illustrates a convolution camera based on the system described by Hosemann and Bagchi (1962). In the original system two identical 'masks' were used consisting of rows of equi-spaced holes. For the purposes of electron microscopic analysis the 'masks' are identical electron micrographs carefully aligned with respect to each other. The image formed by the lens system is the convolution square, N, N−1 of object or 'mask'. (From Elliott et al. 1968.)

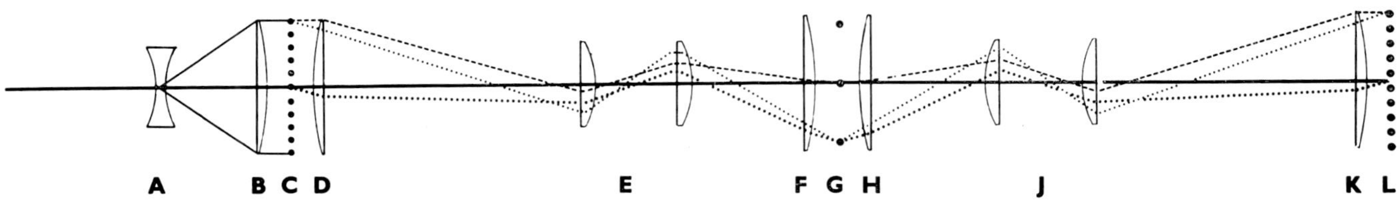

Fig. 7.4. Shows the lens system used in the Polaron diffractometer. Details of the instrument are described in the text. (By kind permission of Polaron Instruments Ltd.)

these components, as many of the basic parts can be made in the average laboratory workshop or purchased from local instrument manafacturers.

Two complete optical diffractometers are presently available commercially and will be described briefly.

7.4.1 POLARON DIFFRACTOMETER

The first is of conventional bench design and is marketed by Polaron Instruments Ltd (see List of suppliers). In order to reduce the length of the diffractometer and image reconstruction system, the optical design is arranged so that the complete instrument occupies only 3 m in length. The basic bench is a precision built double rail giving good rigidity and has a calibrated scale throughout its length for accurate location and reproducibility of the mounted parts. Steel base floor standing units are available which will provide a complete self contained bench if required. The rear and top of the base units are enclosed and darkness is achieved by a curtain system at the front which will allow access to any part of the bench.

The Polaron specifications available to us at the time of writing include those items of the optical system illustrated in Fig. 7.4. Illumination is provided by a helium-neon laser giving a source of coherent light at 1 mW and a wavelength of 632.8 nm. The cross-section of the source from the laser is 2 mm diameter. Adjustable mounts are provided for accurate alignment of the laser beam. Further reduction of the beam diameter is achieved by the insertion of an iris diaphragm which reduces extraneous light from the source. The biconcave lens shown at A, spreads the diameter of the illumination incident on lens B. B is a planoconvex lens which converts the divergent rays to a collimated beam of about 10 mm diameter. The selected electron micrograph is placed in the position C, and the optical diffraction pattern imaged by the planoconvex lens D. A stage of magnification is made possible by the use of a two lens telephoto system comprising the lenses E and F. The enlarged diffraction pattern appears in the plane G, and can be viewed on a fine ground glass screen or recorded on film using a 5 in × 4 in Polaroid Land camera back. Prints from the Polaroid film can be punched to select spots for use as suitable masks for the purposes of image filtration by locating them in a special adjustable holder in the position G, which allows their position to be determined in relation to the original diffraction pattern as described in § 4.4.2.

To reconstruct the image, the mask is placed in position and then the rays passing through the mask are imaged by the telephoto system H and J, which acts as an image reversal system with respect to the lenses E and F. The lens shown at K is a low power projector lens which allows the image to be seen at a suitable magnification at L. Reconstructed images can be recorded by inserting the Polaroid camera in the position L.

For direct enlargement of the diffraction patterns it is possible to remove the lenses H, J and K. The remaining lens system can then be rearranged to make full use of the optical bench length.

7.4.2 TALBOT DESK-TOP DIFFRACTOMETER

The second type of commercial diffractometer at present available is the Talbot PM102 instrument*. This instrument has been designed as a desk-top diffractometer and image processor for specific use with electron micrographs. The dimensions are given as 21″ high, 13″ wide and 18″ deep. The instrument possesses a camera length of 880 mm and the image can be enlarged by X 6 or X 8. The Talbot instrument can also be used as a high quality point source photographic enlarger and can be quickly changed for optical diffraction studies with facilities for image reconstruction.

The optical system in the Talbot diffractometer consists of a 2.4 mW helium-neon laser (Metrologic model MT 211), a beam expanding lens, a pinhole filter, a main diffraction lens and a lens which can be brought into the optical path to focus the diffraction pattern onto the plane of the filtering mask at some distance before the final viewing screen, thus allowing the reconstructed image to be observed on the screen.

The Talbot diffractometer has the advantage of a compact arrangement, achieved by making use of a 'folded' optical system, made possible with the aid of front-gilded mirrors, in contrast to the straight optical bench of conventional diffractometers. An additional advantage of the compact arrangement is that it is possible to scan the micrograph and simultaneously observe the diffraction pattern on the viewing screen.

REFERENCES

Berger, J. E., C. R. Zobel and P. E. Engler (1966), Laser as light source for optical diffractometers: Fourier analysis of electron micrographs, Science *153*, 168.

Elliott, A., J. Lowy and J. M. Squire (1968), Convolution camera to reveal periodicities in electron micrographs, Nature *219*, 1224.

* See List of suppliers.

Elliott, A. and J. Lowy (1970), A model for the coarse structure of paramyosin filaments, J. Mol. Biol. *53*, 181.

Fraser, R. D. B. and G. R. Millward (1970), Image averaging by optical filtering, J. Ultrastruct. Res. *31*, 203.

Gall, J. G. (1967), The light microscope as an optical diffractometer, J. Cell Sci. *2*, 163.

Hosemann, R. and S. N. Bagchi (1962), Direct analysis of matter (North Holland Publ. Co., Amsterdam).

Klug, A. and J. E. Berger (1964), An optical method for the analysis of periodicities in electron micrographs and some observations on the mechanism of negative staining, J. Mol. Biol. *10*, 565.

Lipson, S. G., and H. Lipson (1969), Optical physics (Cambridge University Press).

Squire, J. M. (1971), General model for the structure of all myosin-containing filaments, Nature *233*, 457.

Chapter 8

The enhancement of image detail in electron micrographs by photographic methods

8.1 The principles of photographic averaging applied to electron microscopy

In most physical sciences it is routine practice to attempt to improve the signal-to-noise ratio of measurements by increasing the number of experiments and averaging the results. If the noise or uncertainty is randomly distributed on either side of the mean it sums as its square root, so that the relative contribution of noise to the final result decreases as the square root of the number of observations.

In the case of photographic images the first attempts to use averaging methods of this kind were those of Sir Francis Galton (1879), who was interested in the relationships between the visual appearance of people and such matters as their occupation, state of health and so on. Galton was well aware of the fact that differences between individuals, that may be striking to us, are in reality minor variations about a mean and he combined photographic images of various groups of people in order to see what that mean looked like. The method he used was to produce a composite image on a single photographic plate from a number of photographs superimposed in accurate register with each contribution equally weighted. The results were described by him in the following terms 'the purely individual peculiarities leave little or no visible trace. The latter being necessarily disposed equally on both sides of the average, the outline of the composite is the average of all the components'. If one regards 'individual peculiarities' in the sense of 'noise', then one has an averaging system capable of increasing the signal-to-noise ratio of the final image.

In his work Galton, besides establishing the principle of photographic averaging also encountered and solved the major underlying problems involved. These are: the equal weighting of individual images, the scaling factors required, if any, and the precise registration of the images. The scale was adjusted in this particular instance by using the distance between the eye pupils as the constant scale factor and the two eyes were also used as the fixed registration points. A fourth necessary factor is, of course, the orientation of the head with respect to the line between the eyes. In this particular example it is self evident but in many cases that one may encounter this is not necessarily so.

It has, of course, been obvious for many years that methods of this kind could be employed usefully in electron microscopy, if only because with objects like viruses one encounters hundreds of individual particles, and one would wish to be able to construct an average. In 1958, McLachlan suggested that such methods might be applied to the images of crystalline structures as seen under the electron microscope and he predicted large gains in image resolution. In this he rather overlooked the fact that all one can hope for is an increase of the signal-to-noise ratio to near infinity, and this is not equivalent to an improvement in the signal itself. To take McLachlan's own example, namely a grainy and somewhat out-of-focus picture of a sheet of stamps, the composite has the graininess smoothed out but the out-of-focus image is not appreciably sharpened. This is an important qualification of averaging methods. One can only approach an 'ideal' which itself may be limited by factors other than randomness.

Although, as has been mentioned, averaging methods were known to be desirable, it was not at all evident how they should be carried out. The main problem was establishing datum points and the solution was first obtained for radially symmetrical objects. In these the fixed point is the centre of symmetry, which can be found approximately, and the other factor is the rotation angle which is an integral fraction of 360° and which can be determined by trial and error (Markham et al. 1963). Once the general principles of the procedure had been established such methods were readily applied to objects having linear periodicities and to crystalline and semi-crystalline specimens, not only with the object of reducing random noise, but also to determine hidden periodicities and to select structures having specific periodicities. In this regard photographic integration methods are analogous to tuned filter systems.

8.2 Rotational integration

Markham et al. (1963) described a very simple device for enhancing images having radial symmetry. The apparatus originally used is shown in Fig. 8.1. It consists of a flat board which is placed on the baseboard of a photographic enlarger and has a circular card pivoted on a pin at its centre. The periphery of the card is marked with angular divisions corresponding to integral fractions of 360°, and it also has a set of concentric circles to assist alignment of the centre of the object, or micrograph to be averaged.

The micrograph is placed in the enlarger and projected onto the card. Photographic paper is then attached to the card by adhesive strips (in later work the card was mounted on an iron sheet and the paper held down by button magnets), and the paper given a fraction of the necessary exposure corresponding to the fraction, $1/n$, of 360° being investigated. The card is then turned through the angle $360°/n$ and another exposure made. This is repeated a further $n-2$ times. Maximum reinforcement of detail is given when n is the true periodicity in the object.

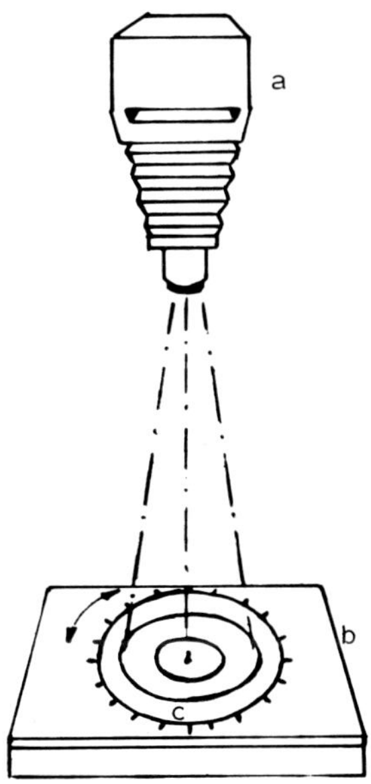

Fig. 8.1. Apparatus for integrating detail in micrographs with radial symmetry. (a) Enlarger unit containing micrograph with region to be studied accurately centred in relation to the lens. (b) Movable board containing rotary disc (c). The centre of rotation must be carefully aligned to the selected region of the micrograph shown at (a).

It will immediately be obvious that any detail on the original will be repeated in the picture n times but, of course, at only $1/n$ the original intensity. Gross flaws in the picture may however give a convincing final image at a spurious periodicity. This was pointed out quite explicitly in the original papers (Markham et al. 1963; Markham 1963) and methods were suggested whereby such artefacts could be detected. Briefly they are: (a) if the periodicity can be factorised reinforcement should also occur at the appropriate fractions of the periodicity, for example if $n = 12$, reinforcement giving a similar final picture should occur at $n = 6$, 4 or 3, and (b) if instead of making n exposures $n-1$ or $n-2$ exposures are given artefact detail will be absent from one or two sectors respectively in the resulting picture, and will easily be detected.

Sometimes objects such as, for example, sections through tubular structures may not be exactly circular and therefore are not always suitable for rotational integration, in which case they may be rendered so by photographically distorting the photographs. Thus a negative of an elliptical object may be made to print adequately circular by projecting it in an enlarger onto a plane that is not parallel to the negative.

The same general technique for obtaining radially integrated electron micrographs was described by Gachet and Thiéry (1964), who applied the method to analyse several biological objects. To allow precise rotation of the images the photographic material was fitted to a rotary milling table. This refinement would appear to be rather superfluous because the limiting factor is in fact the centring of the image rather than precise angular movement. These authors also discussed the possibility of artefact production.

8.3 Rotational integration by stroboscopic methods

Markham et al. (1963) also pointed out that it was possible to approach the problem in other ways. One of them was to mount a photograph on a turntable which was spun by a synchronous electric motor. A stroboscopic light source locked to a multiple of the mains frequency was used to illuminate the image (Fig. 8.2). In this case also the exact centring of the object presented considerable problems.

A refinement, which is technically somewhat exacting was also suggested, and that is spinning the light beam rather than the object itself. This method was suggested independently by Norman (1966) and uses a Dove prism, a right-angled prism, the hypotenuse of which is used to reflect the light beam (Fig. 8.3). When such a prism is rotated the light beam is rotated at twice the prism's rate. Such prisms are used in periscopes and similar optical instru-

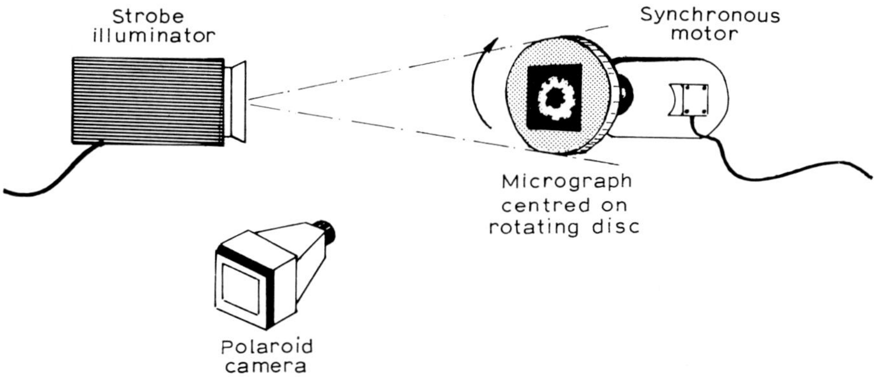

Fig. 8.2. The drawing shows the arrangement of apparatus mentioned in the text for analysing electron micrographs by rotational integration with the aid of a strobe illuminator.

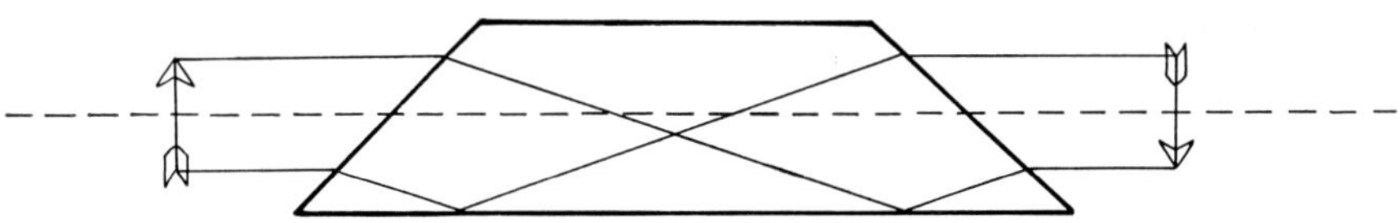

Fig. 8.3. Dove prism showing ray paths. Note that when the prism is rotated about its optical axis (dotted line) the image appears to rotate without any displacement normal to the axis.

ments, but it is difficult to adjust them accurately to spin at high speed exactly about their optic axis. Once this is done, however, the object being examined may be centred at leisure. It is rather doubtful whether this refinement, useful as it is, is justified for ordinary investigation in view of the simplicity and cheapness of the photographic method.

8.4 Linear integration

The applications of photographic averaging techniques to micrographs of objects with linear periodicities presents very substantial problems, unless by chance the periodicity happens to be obvious. In such cases it is only necessary to translate the micrograph in a straight line and photograph its image successively at the appropriate intervals of distance. However, the linear periodicities may not be obvious and in this case an aid to determining rather indefinite periodicities is to superimpose two transparencies of the same micrograph and then to move one in the direction of the suspected periodicity. The beat pattern corresponding to the various periodicities can sometimes be seen. If such attempts fail one must fall back on trial-and-error methods and the apparatus of Markham et al. (1964) was designed with this in view. This apparatus, the linear integrator, which is described in detail in § 8.5, permits photographs to be made in infinitely variable intervals along a chosen direction and in a very short time, so enabling a linear periodicity to be detected even though no clue at all as to its approximate magnitude and direction is available. Nowadays such a search is no longer necessary, since any detectable periodicity may be revealed and its exact spacing determined by optical diffraction methods. It is in conjunction with the diffractometer that the linear integrator is most powerful as a tool. It may also be applied to two dimensional crystalline structures by successive integration along different axes.

8.4.1 CURVE FOLLOWING METHODS

Photographic integration using rigidly defined axes and precise intervals of distance are ideal if the structures to be investigated are almost perfectly crystalline. However, such structures are by no means the general rule in biological material. Rods are frequently not quite straight and they may also exhibit a slight twisting that makes any periodicity run at an angle to the long axis. Curve following methods do not seem to have used so far in electron microscopy and our own attempts to straighten images of rods using anamorphic photography have been unrewarding, mainly because it is very difficult to define the long axes adequately. However, with large two dimensional crystalline objects that have slight defects of this kind it would be feasible to use such methods (see Jenks and Brown 1966, for details of the technique).

8.4.2 IMAGE REINFORCEMENT OF IRREGULAR LATTICE UNIT CELLS

As an alternative it is possible to reinforce the appearance of the unit cell itself in a slightly irregular lattice by registering the cells individually and photographing them in succession. To do this it is necessary to have some sort of guide as to lattice spacing and orientation. In this event it is desirable to draw coordinates on tracing paper using red ink and to use these for orientating the individual unit cells in the lattice, which has been enlarged onto a sheet of cut film. If a red filter (Wratten A) is used over the camera lens the coordinates will not appear in the final averaged picture of the unit cell. In this case a certain subjective element enters, but this may be reduced somewhat by going over the lattice several times in making up the composite picture.

8.4.3 'SYMMETRISING' VIRUS PARTICLES

A simple example of this type of approach is given by the 'symmetrising' of a virus particle. 'Spherical' virus particles frequently come to rest on their two-fold axes and in this position, not only is the back of the particle in register with the front, giving a reinforcement of structure, a composite picture in which detail is present twice, but the details are repeated three more times due to the reflection planes (Fig. 8.4). Therefore a further improvement may

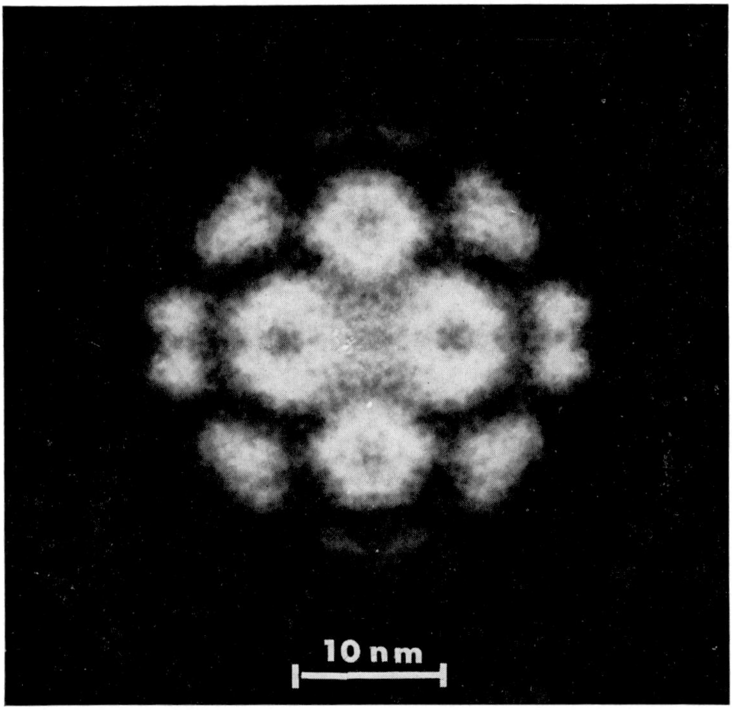

Fig. 8.4. The electron micrograph shows a particle of turnip yellow mosaic virus after being 'symmetrized' about its twofold position. The image was obtained by superimposing four views of the virus. A first exposure was made, and then the micrograph was rotated through 180° in the enlarger, and a second exposure made. The micrograph negative was then reversed by turning it over, and a third exposure made. This was followed by a fourth exposure, after another rotation of 180°. All of these views are equivalent because the particle shown has two mirror planes at right angles to each other.

be achieved by printing the photograph four times rotating it through 180° each time i.e. rotating in the plane of the image and turning the transparency over and rotating again. The guide lines are the mirror planes and the image is centred on these by eye. In order to avoid a starkly rigid image each of the four positions is positioned at least five times. This has the additional advantage of minimising photographic grain (Fig. 8.4). This procedure is not applicable to the 'skew' series of icosahedral viruses. (See Horne and Wildy 1961; Caspar and Klug 1962).

8.4.4 IMAGE REINFORCEMENT OF TWO-DIMENSIONAL REGULAR LATTICE PATTERNS

Another example of this type of linear integration is provided by an electron micrograph with a two-dimensional lattice structure obtained from part of a gold foil specimen which was recorded at high magnification. The spacings and angles of the repeating features were obvious and two-dimensional superimposition by direct photography was a relatively simple procedure. The method is illustrated in Fig. 8.5, and consists of illuminating an enlarged transparency of the original micrograph to be integrated. In order to move the transparency across the illuminating box at the correct intervals as well as for alignment purposes, a strip of the transparency was cut and fixed to the box as indicated. The calculation for exposure was the same as mentioned earlier by dividing the total exposure required by the number of integrations to be made. Integration was carried out by moving the larger film across the box at the correct intervals and using the smaller film as a reference to determine the overlap distance between the two. The electron micrograph (Fig. 8.6) shows the result of integration over a square of 81 intervals (9 by 9 atom diameters).

8.5 Design, construction and operation of a wide range linear integrator

In considering the basic design requirements for an apparatus for the superimposition of images of linear objects Markham et al. (1964) took into account the fact that movement in the direction of integration had to be precise and that the periodic features were required to be orientated at the correct angle in relation to the direction of integration. Moreover, the number of integrations carried out by the apparatus had to be infinitely variable. To meet these demands the apparatus illustrated in Fig. 8.7 was constructed; it is a modified version of the system described in the original publication. Basically, the apparatus consists of a rectangular table constructed from

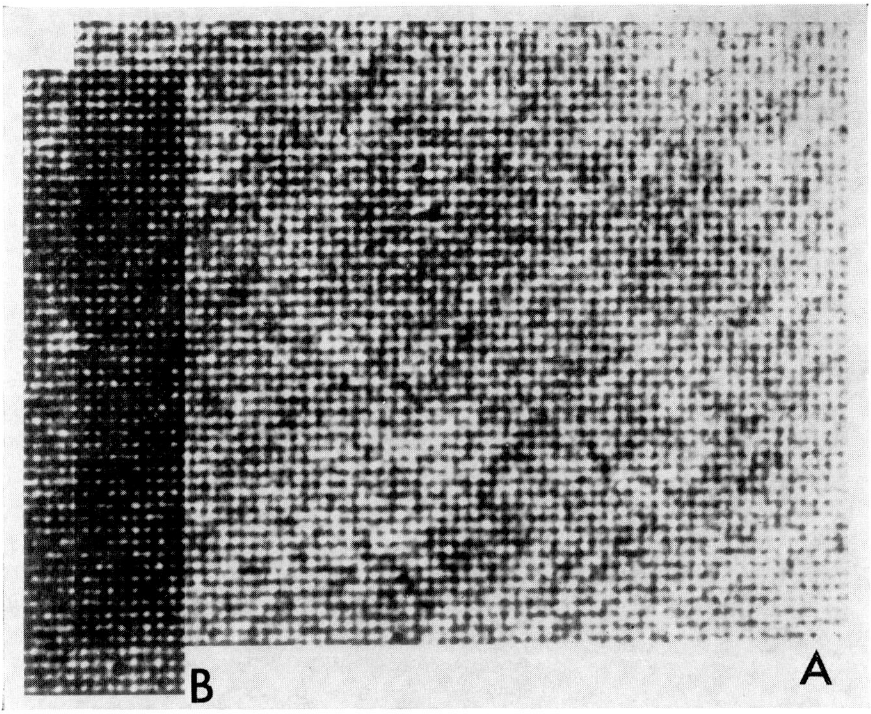

Fig. 8.5. A simple method for the reinforcement of a two-dimensional regular lattice. The enlarged transparency shown at A was obtained from an original electron micrograph and placed on an illuminator box. Integration is achieved by moving the transparency N intervals in one direction and N_1 intervals in the other. The strip of film shown at B was cut from A and used as a reference for the accurate location of the transparency between exposures. The result of the procedure is shown in Fig. 8.6.

wood blockboard A, with an aperture cut as shown at B. Attached to the underside of the table are two metal rod runners C, which are only in contact with the V-blocks D which act as guides for A. The V-blocks are made from teflon or similar material to reduce friction. The white card indicated at E is used as a triggering scale. A series of pencil lines drawn at regular intervals in a fan-shaped pattern allow very fine adjustments to be made at right angles to the direction of the table movement. The stroboscopic lamp, F, located in the position shown is triggered by a pick-up device N, which makes contact with the pencil lines on E, thus operating an electronic relay G, when the table is moved in the direction indicated by the arrows. The

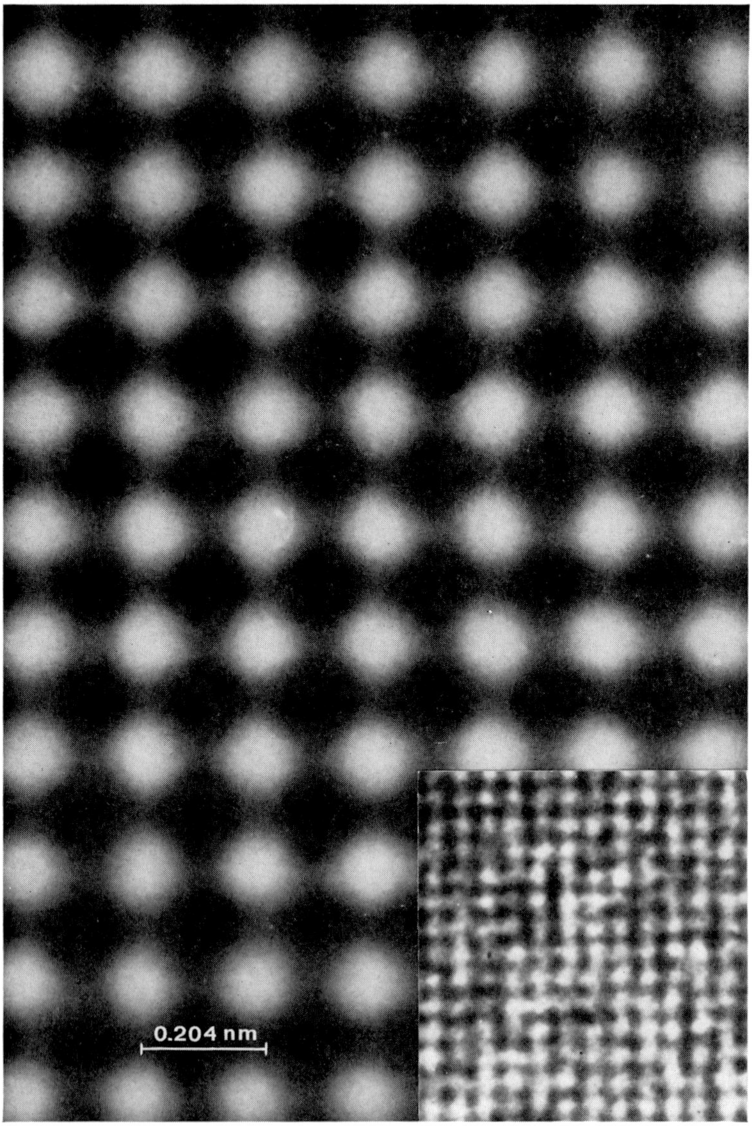

N

N_1

Fig. 8.6. Atomic lattice of gold. The photograph was obtained by the application of the simple photographic method illustrated in Fig. 8.5. The photograph was the result of averaging, $N = 9$ intervals and $N_1 = 9$ intervals. Part of the original micrograph resolving the lattice structure in a thin gold foil is shown at lower magnification in the insert. The actual shape of the atoms, which are in white, is only approximate, the overall resolution not exceeding 0.01 nm. See also the optical diffraction pattern in Fig. 6.3 which was taken from the original electron micrograph.

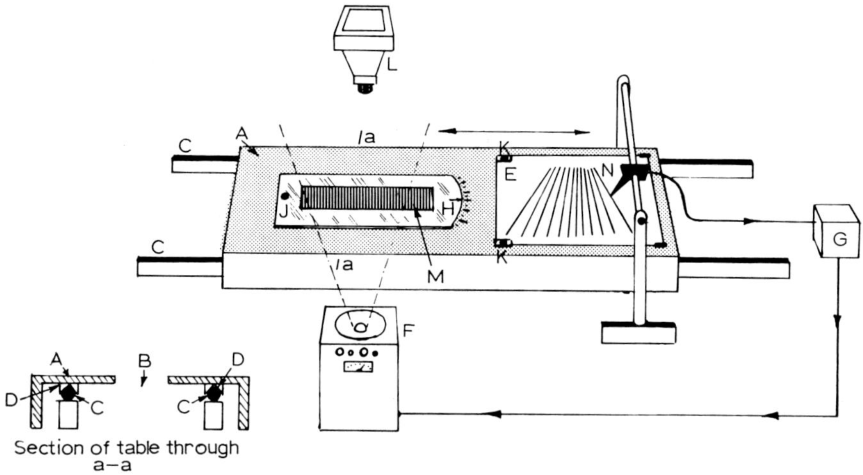

Fig. 8.7. The basic arrangement of the linear integrator described in the text is shown in diagrammatic form. (From Markham et al. 1964.)

pick-up can be moved along the pencil lines by means of the screw thread M, for fine adjustment of the integration interval.

The procedure for linear integration of an image using this apparatus is as follows: an enlargement of the area to be integrated is made directly onto film of high contrast, the extent of enlargement obviously depending on the estimated spacings present in a given object. Small structures such as rods of tobacco mosaic virus may require a final magnification on the film corresponding to several million times, whereas sperm tails may need far less enlargement because the object and its structural spacings are much larger. In addition, the object must be relatively straight along the axis to be integrated.

The micrograph transparency is fixed in position on the perspex protractor, H, over the aperture, B, and the correct angle of integration set by trial and error by moving the protractor about the position J. A suitable trigger card is selected possessing the estimated range of triggering intervals and placed in the clamps K. By placing the pick-up device N on the space in front of the first pencil line on E, and moving the table smoothly in the direction indicated, each pencil line will trigger the strobe lamp and the individual stroboscopic flashes will illuminate the micrograph intermittently at the triggering intervals. The integrated image is recorded by the camera, L, a Polaroid Land camera back being used for rapid recording. Some indication of the quality and accuracy of the final image can be judged by the integrated photographs of a platinum phthalocyanine crystal lattice shown in Fig. 8.8.

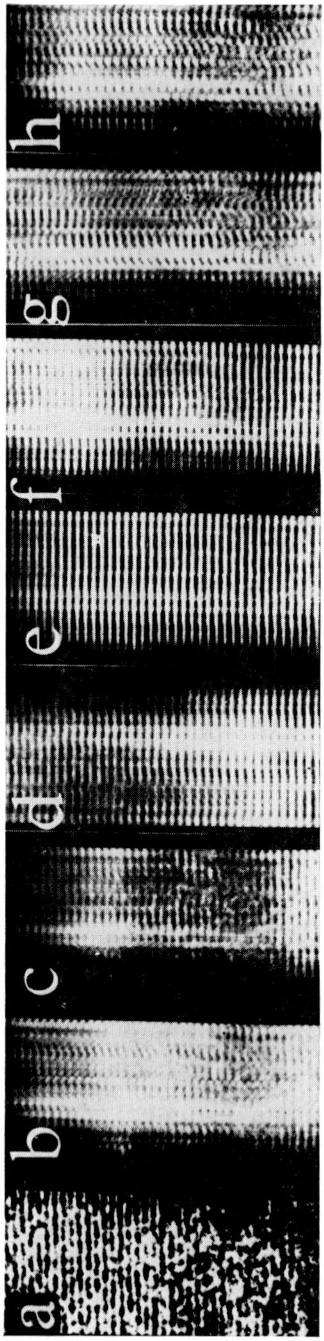

Fig. 8.8. Electron micrographs showing linear repetitive features subjected to integration with the aid of the apparatus shown in Fig. 8.7. The lattice spacings in Pt phthlocyanine crystals serve as a good illustration for the application of this technique to provide image reinforcement and accurate measurement. The series of photographs shows the original image (a) after integration at periodicities of b = 1.116, c = 1.142, d = 1.168, e = 1.194, f = 1.220, g = 1.246 and h = 1.272 nm respectively. (From Markham et al. 1964.)

One of the difficult problems the user has to face is the determination of the approximate spacings and angles present in the original image for the purposes of integration. As mentioned earlier, the most obvious way to do this is with the aid of the optical diffractometer which essentially presents the same information, but in reciprocal lattice space. An example of how the techniques of optical diffraction and photographic integration can be combined is well illustrated in the publication of Hitchborn and Hills (1968).

8.6 Simple 'shadow' linear integrator

Warren and Hicks (1971) have recently devised a simple and rapid method for obtaining the linear integration of electron micrographs. The method has the advantage of direct visualization over those methods mentioned earlier where it is often necessary to obtain the correct spacings by trial and error. They applied the principle that when a micrograph transparency possessing periodic features is illuminated with several point sources of light the resulting shadow cast will have the same density distribution as the micrograph. The geometrical arrangement of the system is shown in Fig. 8.9. By placing the micrograph between a series of equally-spaced point light sources and a screen, image averaging can be achieved by adjusting the positions of the

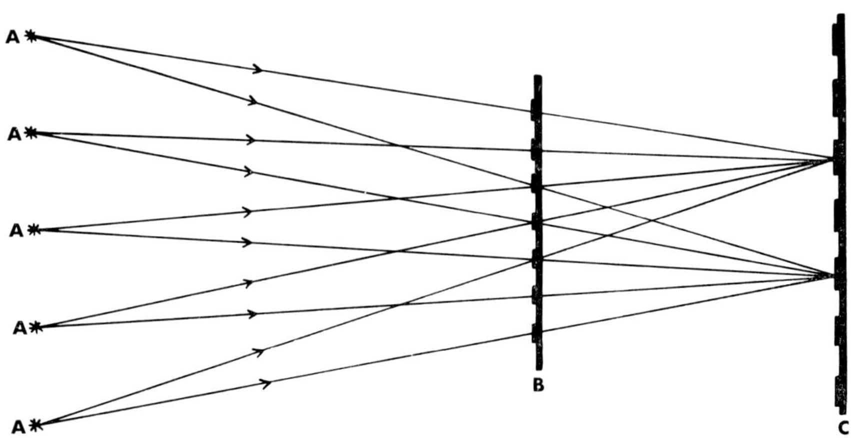

Fig. 8.9. The diagram illustrates the basic geometry of the shadow integrator. The light rays from the point sources (A) pass through the enlarged transparency (B) obtained from the original electron micrograph and the overlapping shadows are cast on to the screen (C). Shadows of the periodic features present in the transparency fall in register in the plane of the viewing screen with some degree of magnification. (From Warren and Hicks 1971.)

screen, light sources or micrograph. Warren and Hicks (1971) showed that if n is the integration interval

$$\frac{\text{Image resolution}}{\text{Image periodicity}} = n \times \frac{\text{Source diameter}}{\text{Source separation}}.$$

This information is independent of the positions of the micrograph and screen. In order to obtain maximum resolution from the final integrated image, the equal distances separating the light sources should be as large as possible within practical limits. The size of the light sources on the other hand, should be small and the lowest value of n should be applied.

8.6.1 DESIGN AND CONSTRUCTION OF A 'SHADOW' LINEAR INTEGRATOR

The materials required to construct the apparatus are easy to obtain, low in cost and require the minimum amount of machining or workshop equipment. The general arrangement of the apparatus is illustrated in the diagram Fig. 8.10 which shows the relative positions of the parts. It can be seen that apart from the Polaroid-Land camera, the components are housed inside a light-tight box. The box can be constructed from standard Dexion angle to form an inner frame which is then covered with hardboard. In order to reduce reflections and scattered light the interior of the box is painted with a matt black finish.

Twenty 6.5 V 0.3 A lamps are mounted in standard holders which are fixed to a batten and wired in parallel. The centre-to-centre distance (5 cm) between the holders should be reasonably accurate and the filaments of the lamps aligned in the vertical position. In practice the location of the filaments within their glass bulbs is frequently found to be off-centre and the allowance for some minor adjustment of the individual bulb holders may be advantageous; they are then secured by the clamping screws as shown in Fig. 8.11. To prevent a voltage drop across the electrical system and consequent uneven illumination, the row of lamps is fed from both ends from a 6 V 5 A transformer. Control over the intensity of the lamp system can be achieved by inserting a suitable variable resistor or alternatively with the aid of a Variac to control the input supply to the transformer.

From Fig. 8.9 it can be seen that focussing is achieved by moving the micrograph between the fixed positions of the light source and camera or viewing screen. The micrograph is mounted between two Perspex or Plexiglass sheets as illustrated in Fig. 8.12. Two mechanical movements of the micrograph are required; firstly, it must be capable of being supported in the

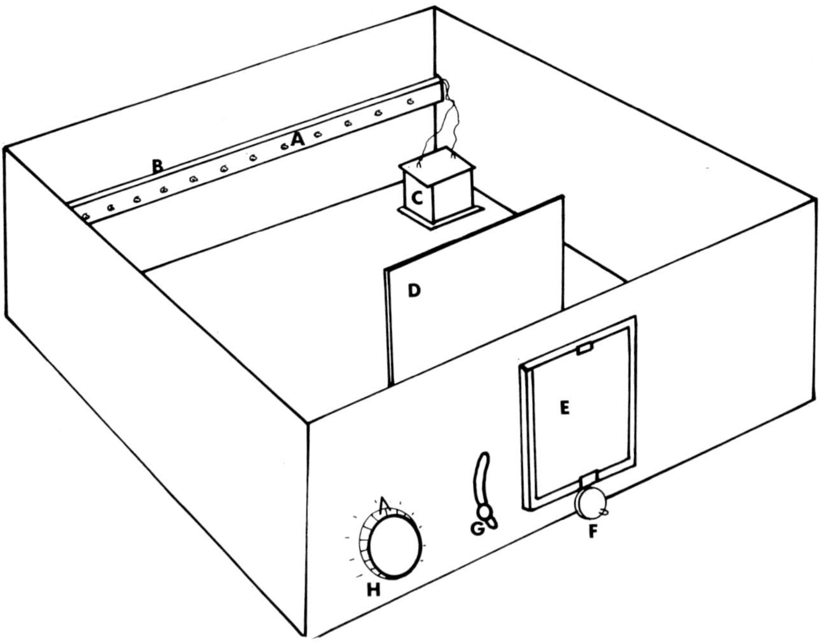

Fig. 8.10. The drawing illustrates the general arrangement of the components forming the shadow integrator described by Warren and Hicks (1971). The illuminating source consists of a series of low voltage bulbs (A), mounted on a strip (B), and supplied from the transformer (C). The enlarged transparency from an electron micrograph is placed between the Perspex sheets (D). The image is viewed on the screen (E), or photographed by replacing the screen with a Polaroid 5 × 4 camera back. Linear adjustment of the micrograph between the screen and illuminating source is controlled by the knob (F), and rotation by the lever (G). The control (H), is used to set the level of light intensity by controlling the input voltage to the transformer (C). For constructional details of the transparency holder and illumination strip see Fig. 8.11 and Fig. 8.12. (From Warren and Hicks 1971.)

upright position and moved accurately between the light source and screen, secondly some provision has to be made for tilting it in relation to the line of bulbs. A simple frame device of the type shown in Fig. 8.12, meets the basic mechanical requirements. The threaded rod which drives the micrograph holder in a horizontal direction and the tilting handle are located at the front of the box.

8.6.2 PHOTOGRAPHIC AND CONTRAST REQUIREMENTS FOR LINEAR INTEGRATION

This integrating process produces a loss in contrast and for this reason Warren and Hicks (1971) have suggested the use of Kodalith Ortho Type 3

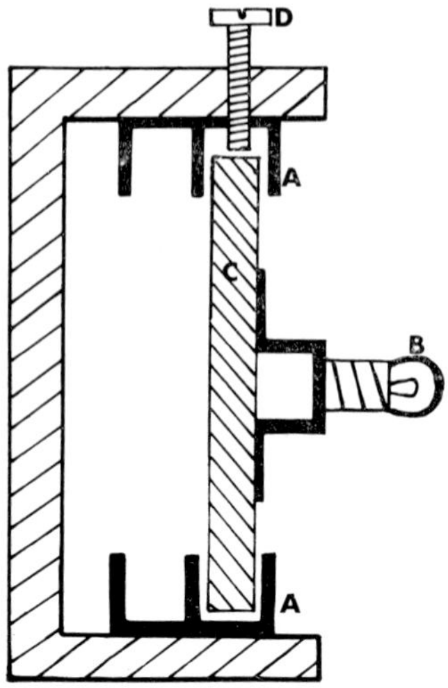

Fig. 8.11. The drawing shows a cross-section through the bulb mounting strip (A) fitted to the shadow integrator. Each bulb (B) is located on a sliding mount (C), for accurate spacing in relation to its neighbouring bulb. Once in position the bulbs are locked with the aid of the securing screw (D).

film, developed in Kodalith Liquid developer for making the micrograph transparencies. There is the problem of density gradient amplification associated with high contrast films of this type and some method of compensating these effects may be necessary. In their original publication Warren and Hicks recommended the production of a mask by making a preliminary exposure on film giving approximately equal areas of over and under exposed emulsion. This first film was used as a mask and placed over the second film during exposure, the first film being moved irregularly in relation to the second film for all or part of the duration of the second exposure. This enabled the lighter parts of the image to be suppressed by the darker parts of the exposed film, but the darker regions of the image remained unaffected. Movement of the mask in an irregular manner allowed small scale density differences to remain unaltered. When making the enlargements for integration with this apparatus the range of periodicities has to be suitably adjusted so that the first order spacings observable in the integrator set up as described above are within the range of 0.15 cm to 2 cm.

The enlarged micrograph is placed between the two Perspex plates as shown in Fig. 8.12 and the integrated image recorded on suitable Polaroid film.

In addition to being able to observe the integrated image directly on a viewing screen, an image possessing two-dimensional lattice features can be recorded on the Polaroid film, printed onto Kodalith film following the same masking process mentioned earlier and then re-integrated along a

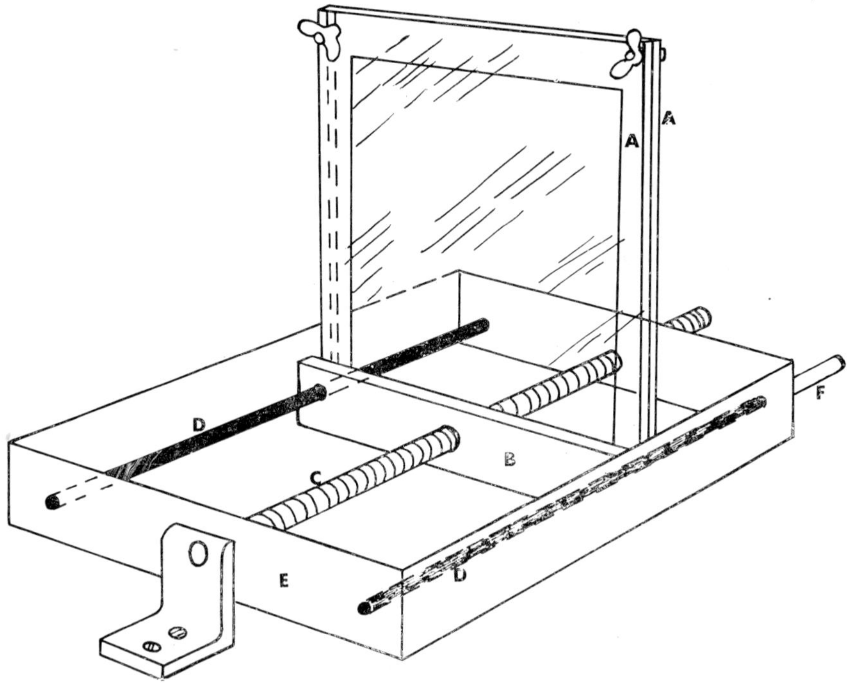

Fig. 8.12. The sketch shows the basic constructional details of the micrograph transparency holder in the shadow integrator. Two Perspex sheets (A), are mounted to the sliding bar (B). The lead screw (C), allows the transparency holder (A), to be moved in the desired direction along the guiding rods (D). The frame (E), is tilted about the axis of (C), using the control (F). Alignment of the lead screw (C), with respect to the guiding rods (D), is critical in order to allow smooth operation.

different direction. The subunits present in rat bladder luminal membranes are illustrated in Fig. 8.13 and are a good example of the application of the method to a biological object. Information concerning the spacings and angles of the subunits were intially determined by optical diffraction. (See Vergara et al. 1969, and Warren and Hicks 1970.)

8.7 Enhancement of images by cylindrical lenses

A technique for the enhancement of the signal-to-noise ratio in electron micrographs was briefly discussed by Dowell et al. (1968), and was applied to micrographs which contained banded patterns. The micrograph was placed in a photographic enlarger and projected in the normal way. A rotat-

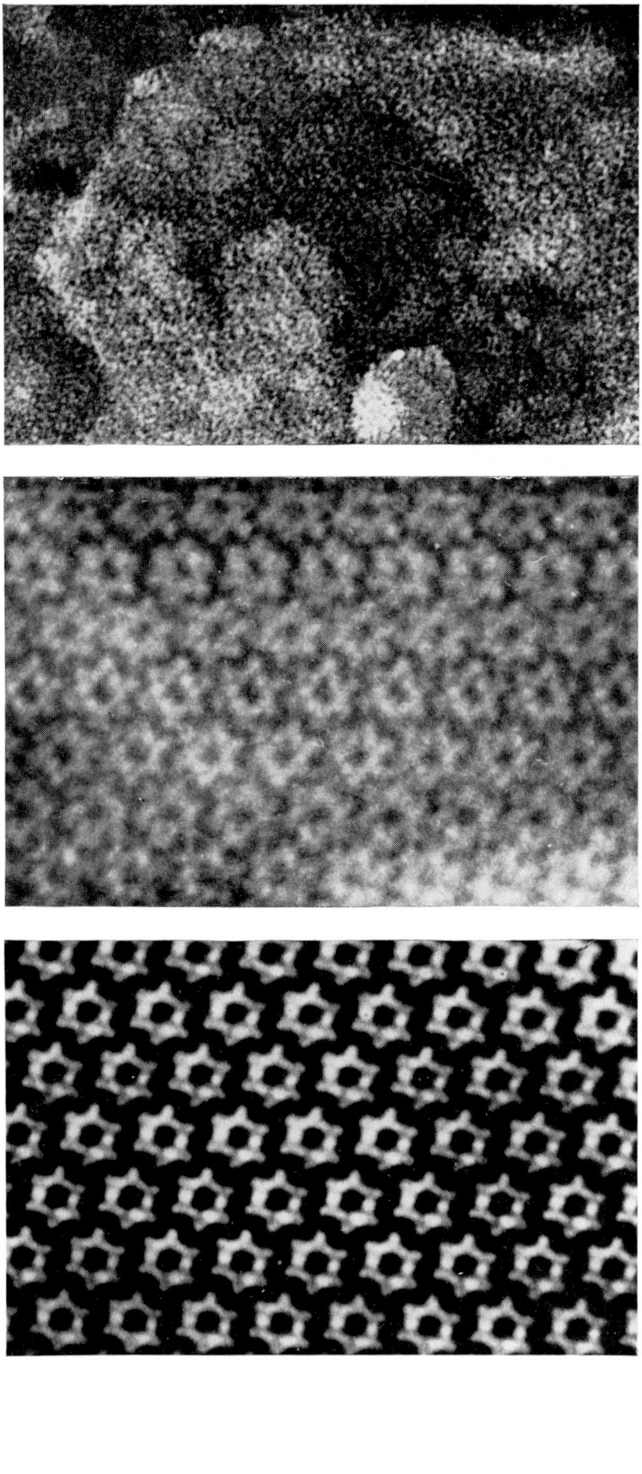

(A)

(B)

(C)

able cylindrical lens was fitted immediately below the enlarger lens with the result that the combined lens system gave a line image from normal object points. It was pointed out that in a plane near the lenses the lines are parallel to the axis of the cylindrical lens and in a remote plane the lines are at right angles. The remote plane was considered to be more convenient to use as in this plane the magnification is the same as that of the image formed by the enlarger lens alone. It was observed that when the image from a banded structure was formed with the aid of this optical system there was a considerable improvement in the signal-to-noise ratio. It must be emphasized that such a system is only capable of giving enhanced information in one direction and that any periodicities within the band patterns themselves will be lost. This point was well illustrated in the published electron micrographs of crystallites of calf collagen where the main band pattern densities appearing across the crystallite axis were considerably reinforced, but some of the small repeating detail which was seen across the axis in the normal micrograph was removed. This method is analogous to placing a mask in the form of a slit in front of a photographic film and moving the micrograph across the slit in the direction of the banded structures.

A second method suggested by Dowell et al. (1968) merely involved placing a mask in the plane of an optical diffraction pattern formed by a special projector lens and thus absorbing part of the zero order pattern.

REFERENCES

Caspar, D. L. D. and A. Klug (1962), Physical principles in the construction of regular viruses, Cold Spring Harbor Symp. Quant. Biol. *27*, 1.

Dowell, W. C. T., J. L. Farrant and J. D. McLean (1968), The enhancement of the signal-to-noise ratio in micrographs, in: Fourth European Regional Conference on Electron Microscopy. Vol. 1, D. Steve Bocciarelli, ed. (Tipografia Poliglotta Vaticana, Rome), p. 583.

Gachet, J. and J. P. Thiéry (1964), Application de la méthode de tirage photographique avec rotations ou translations à l'étude de macromolecules et de structures biologiques, J. Microscopie *3*, 253.

Galton, F. (1878), Composite portraits, Nature *18*, 97.

Fig. 8.13. Shadow line integration. The original electron micrograph from a preparation of rat bladder luminal membrane is shown at A. There is only a slight indication of a possible regular pattern. With the aid of the apparatus shown in Fig. 8.10 it was possible to obtain a shadow line integrated image as shown in B which is the result of the first integration. The second image shown at C results from reintegrating B at an angle of 60° to the first integration. The centre-to-centre distance of the pseudo-hexagonal array is about 1.4 nm. (Reproduced by kind permission of Dr. R. Warren and Dr. Marion Hicks.)

Hitchborn, J. H. and G. J. Hills (1968), A study of tubes produced in plants infected with a strain of turnip yellow mosaic virus, Virology *35*, 50.

Horne, R. W. and P. Wildy (1961), Symmetry in virus architecture, Virology *15*, 348.

Jenks, G. F. and D. A. Brown (1966), Three-dimensional map construction, Science *154*, 857.

Markham, R. (1963), Substances produced during replication of plant viruses, in: Viruses, nucleic acids and cancer, Seventeenth Annual Symposium on Fundamental Cancer Research (Williams and Wilkins, Baltimore), p. 180.

Markham, R., S. Frey and G. J. Hills (1963), Methods for the enhancement of image detail and accentuation of structure in electron microscopy, Virology *20*, 88.

Markham, R., J. H. Hitchborn, G. J. Hills and S. Frey (1964), The anatomy of the tobacco mosaic virus, Virology *22*, 342.

McLachlan, D. (1958), Crystal structure and information theory, Proc. Nat. Acad. Sci. U.S. *44*, 948.

Norman, R. S. (1966), Rotation technique in radially symmetric electron micrographs: mathematical analysis, Science *152*, 1238.

Vergara, J., W. Longley and J. D. Robertson (1969), A hexagonal arrangement of sub-units in membrane of mouse urinary bladder, J. Mol. Biol. *46*, 593.

Warren, R. C. and R. M. Hicks (1970), Structure of the subunits in the thick luminal membrane of rat urinary bladder, Nature *227*, 280.

Warren, R. C. and R. M. Hicks (1971), A simple method of linear integration for resolving structure in periodic lattices: application to an animal cell membrane and a crystalline inclusion, J. Ultrastruct. Res. *36*, 861.

Chapter 9

Computer processing of images and diffraction patterns

9.1 Mathematical analysis and processing of diffraction patterns

It is, of course, possible to analyse the optical diffraction pattern mathematically, and to compensate at this stage for certain discrepancies that may have arisen in the formation of the diffraction pattern. The amplitudes have to be deduced from the spot intensities, which is simple in the case of centrosymmetrical objects having a strong scatter at the centre (Lipson and Lipson, 1969, p. 276) in which all the spots happen to be in phase. Objects not having this desirable attribute can be processed in such a way as to acquire it by printing two images rotated by 180° on the plate and with a large clear area in the centre. It may be necessary to attenuate the two images, in which case fine gauzes can be used. This procedure has been used by Wyckoff et al. (1957) for optical Fourier synthesis, which is, of course, similar to the process of image reconstruction, but in this case starting from the plane of the diffraction pattern rather than from the plane of the image.

It is also possible, in theory at least, to determine phase relationships in the plane of the diffraction pattern by superimposing a reference signal from the light source. One may also measure the intensity distribution in the original micrograph, calculate the diffraction pattern, modify it mathematically and finally reproduce the modified micrograph.

The optical diffractometer analyses micrographs by transforming the distances between elements in the micrograph and their relative orientations into vectors in reciprocal space. In addition modulation in the original is presented as differences in the amplitudes of the various spectral orders in the optical diffraction patterns. Image reconstruction techniques reform the image by repeating this process and so recover the original picture. It is possible to modify the elements in the diffraction pattern before reconstruction to form an altered and possibly improved version of the original.

It is obvious that, since the optical system carries out mathematical operations, and indeed is itself an optical analogue computer, it should be possible to perform similar operations by means of digital computers (see Nathan 1971; Billingsley 1971). In order to do this the original micrograph has to be analysed into its elements and the densities and relative orientations of these measured and recorded. This can be done automatically by means of automatic densitometers, such as the Joyce-Loebl Autodensitometer (see List of Suppliers) and others.

Having anlysed the picture into elements, one may compute a number of functions, such as diffraction patterns or Fourier transforms, (Andrews, 1970) introducing weighting factors as required and finally recover the image with the contrast adjusted, noise removed, and so on. Much of this work has been done on rare pictures taken in outer space, having imperfections and so far relatively little has been done on electron micrographs, and most of that on micrographs selected without much discrimination. The major advantages in computer processing is the relative ease with which it is possible to modify the amplitudes of selected spatial frequencies by means of mathematical filters of any desired characteristic and even to change phases if desired. The disadvantage, apart from relative cost, is that the analysis of the pictures into finite areas may reduce the amount of information and may also introduce spurious detail associated with the sampling intervals. Lens systems do not have this defect in that resolution of picture elements is limited only by diffraction, but the filtering required in this case is much more involved and is limited to simple linear modifications of amplitudes. Phases may also be adjusted but this procedure requires the production of special phase-plates for each particular picture. Stroke (1971) has recently described in detail a system for accomplishing simple amplitude adjustment and phase changing using holographic methods. The results obtained in this way do not suggest that as much improvement may be obtained as one might hope for. This may be associated with the optical problems of producing 'ideal' holograms.

9.2 Image averaging by computer methods

Computer image processing may also be applied to image averaging methods. In this case the computer may be programmed to detect and display rota-

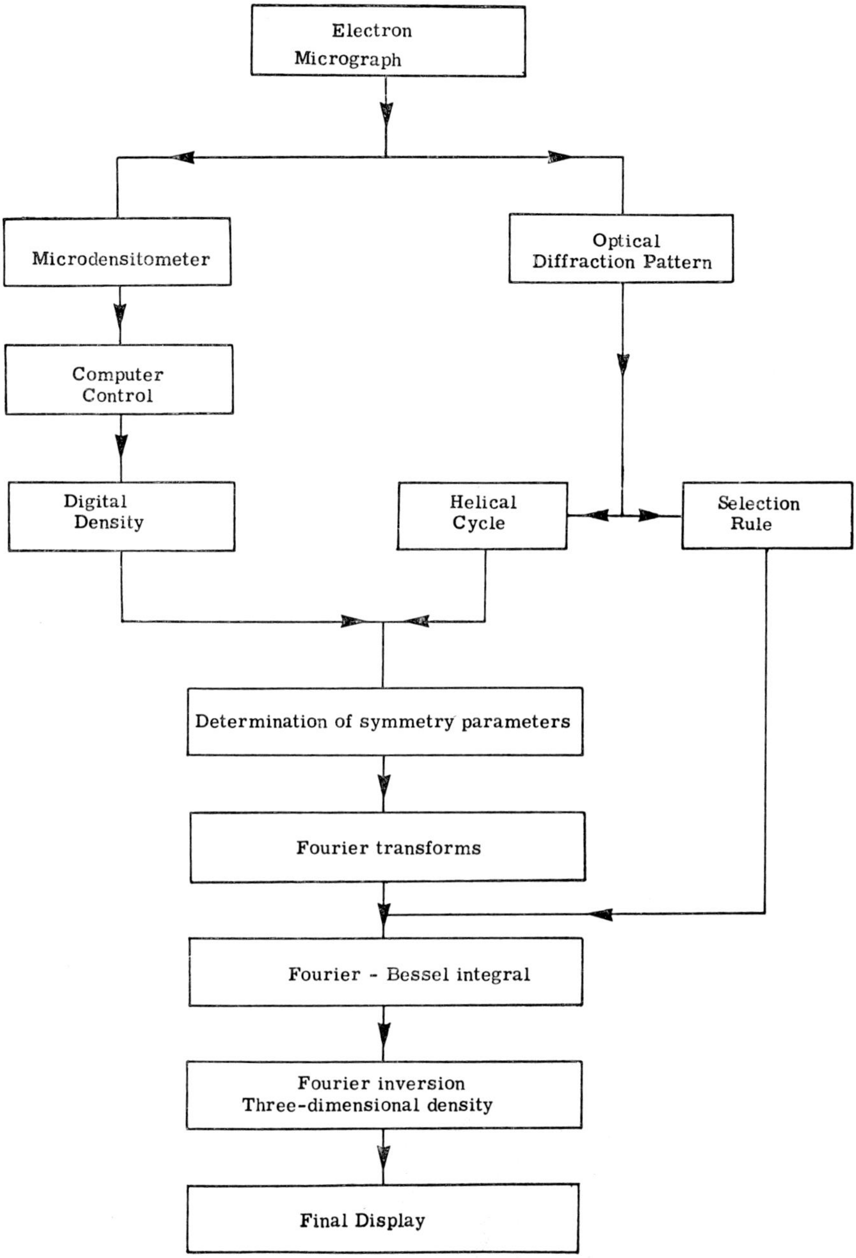

Fig. 9.1. Block diagram showing the various stages required for image analysis by computer and data processing methods.

tional symmetry elements, the centre of symmetry and the azimuth orientation of the object (Crowther and Amos 1971). This type of method has considerable potentialities in that multiple image superimposition may be carried out for a number of averaged objects. A difficulty associated with this is the occurrence of left and right-handed objects having similar symmetries. Averaging would, of course, necessitate selecting each type unequivocally.

9.3 Three-dimensional image processing by computer methods

Computer image processing of a more complex type and somewhat outside the scope of this book is aimed at the reconstruction of three-dimensional images from the analysis of a number of random two-dimensional pictures of an object taken in different orientations (De Rosier and Klug 1968). The simple reasoning behind this reconstruction method is that, for example, a helix is its own stereoscopic pair if it is displaced along its axis by the appropriate amount. Similarly any two photographs of an icosahedron other than two identical ones along two-fold axes can in theory be orientated to form a stereoscopic pair. Objects having less symmetry have to be examined in a greater number of orientations but, of course, they must all be identical. In these methods it is difficult to take full advantage of the redundancy available in many electron micrographs, but at least in theory one should be able to effect a considerable improvement once the methodology has been conquered. The diagram shown in Fig. 9.1 illustrates the various stages used in processing electron micrographs by computer methods.

REFERENCES

Andrews, H. C. (1970), Computer techniques in image processing (Academic Press, New York).

Billingsley, F. C. (1971), Image processing for electron microscopy: II. A digital system, in: Advances in optical and electron microscopy, Vol. 4, R. Barer and V. E. Cosslett, eds. (Academic Press, New York), p. 127.

Crowther, R. A. and L. A. Amos (1971), Harmonic analysis of electron microscope images with rotational symmetry, J. Mol. Biol. *60*, 123.

De Rosier, D. J. and Klug, A. (1968), Reconstruction of three-dimensional structures from electron micrographs, Nature *217*, 130.

Lipson, S. G. and H. Lipson (1969), Optical physics (Cambridge University Press).

Nathan, R. (1971), Image processing for electron microscopy: I. Enhancement procedures. in: Advances in optical and electron microscopy, Vol. 4, R. Barer and V. E. Cosslett, eds. (Academic Press, New York), p. 85.

Stroke, G. W. (1971), Sharpening images by holography, New Scientist *51*, 671.

Wyckoff, H. W., R. S. Bear, R. S. Morgan and D. Carlstrom (1957), Optical diffractometer for facilitation of X-ray diffraction studies of macromolecular structure, J. Opt. Soc. Am. *47*, 1061.

List of suppliers

The following is a list of suppliers presently known to the authors. They would be pleased to receive notification of any relevant additions to the list.

General suppliers of electron microscope accessories and of chemicals and metals in Great Britain and the U.S.A. are listed first, from whom the majority of small items required may be obtained.

General electron microscope accessory suppliers

Aeon Laboratories,
Beech Hill,
Ridgemead Road,
Englefield Green,
Egham,
Surrey, U.K.

Ernest Fullam Inc.,
P.O.Box 444,
Schenectady,
New York 12301, U.S.A.

Ladd Research Industries Inc.,
P.O. Box 901,
Burlington,
Vermont 05401, U.S.A.

Polaron Equipment Ltd.,
4 Shakespeare Road,
Finchley, London N. 3, U.K.

or

Polaron Instruments Ltd.,
60/62 Greenhill Crescent,
Holywell Estate,
Watford,
Hertfordshire, U.K.

TAAB Laboratories,
52 Kidmore End Road,
Elmer Green,
Reading,
Berkshire, U.K.

Polysciences Inc.,
Paul Valley Industrial Estate,
Warrington,
Pennsylvania 18976, U.S.A.

General chemical and metal wire suppliers

BDH Chemicals Ltd.,
Poole,
Dorset, U.K.

Hopkin and Williams,
Freshwater Road,
Chadwell Heath,
Essex, U.K.

Johnson, Matthey and Co. Ltd.,
73/83 Hatton Garden,
London EC1, U.K.

Koch-Light Laboratories,
Colnbrook,
Buckinghamshire, U.K.

Specific suppliers

1 *Aluminium wire*

0.75 diameter

BDH Chemicals Ltd.

0.015″ (15 mil) diameter

Ernest Fullam Inc.

Ladd Research Industries Inc.

0.028″ and 0.048″ diameter

Hopkin and Williams

Range of diameters

Johnson, Matthey and Co. Ltd.

1 mm diameter

Koch-Light Laboratories Ltd.

0.25 mm diameter

Polaron Equipment Ltd.

2 *Aquadag. – colloidal graphite*

Acheson Colloids Ltd.,
P.O. Box 12,
Prince Rock,
Plymouth,
Devon, U.K.

2a *Autodensitometer*

Joyce-Loebl and Co. Ltd.
Princeway,
Team Valley,
Gateshead,
Co. Durham, NE 11 0UJ
England

3 *Automatic print processors*

Agfa-Gevaert Ltd.,
Great West Road,
Brentford,
Middlesex, U.K.

Ilford Ltd.,
Ilford,
Essex, U.K.

Kodak Ltd.,
P.O. Box 33,
Hemel Hempstead,
Hertfordshire, U.K.

or local suppliers

4 *Catalase*

Boehringer Mannheim GmbH,
Mannheim,
Germany
Biochemical Department

The Boehringer Corporation (London) Ltd.,
Bilton House,
54/58 Uxbridge Road,
Ealing, London W5 2TZ, U.K.

Polaron Equipment Ltd.

5 *Crystal and reciprocal lattice models*
Crystal Structures,
Bottisham,
Cambridge, U.K.

6 *Diffraction calibration specimens*

a *Aluminium*

Ernest Fullam Inc.,

Ladd Research Industries

b *Thallous chloride*

Ernest Fullam Inc.

7 *Diffraction gratings*

Mason and Morton Ltd.,
Fir Tree House,
Headstone Drive,
Wealdstone, Harrow,
Middlesex, HA3 5QS, U.K.

Polaron Instruments Ltd.

Ernest Fullam Inc.

8 *Filters*

Sir Howard Grubb Parsons and Co. Ltd.,
Shields Road,
Walkergate,
Newcastle-upon-Tyne, NE6 2YB,
U.K.

Mason and Morton Ltd.

Polaron Instruments Ltd.

Ernest Fullam Inc.

9 *Gold Wire*

0.008″ diameter

Ernest Fullam Inc.

1 mm diameter

Koch-Light Laboratories Ltd.

Range of diameters

Johnson, Matthey and Co. Ltd.

0.008″ diameter

Ladd Research Industries Inc.

0.010″ diameter

Polaron Equipment Ltd.

10 *Laser and Laser optical accessories*

C.W. Radiation Inc., *represented by* Rofin Ltd.,
3 Windhill,
Bishop's Stortford,
Hertfordshire, U.K.

Spectra Physics,
5 Wolsey Road,
Hemel Hempstead,
Hertfordshire, U.K.

Scientia and Cooke Electronics Ltd.,
40–48 High Street,
Acton, London W3, U.K.

11 *Lenses, lens systems, optical benches, lens holders and mounts*

George Elliot and Sons Ltd.,
Worcester House,
Vintners Place,
London EC4V 3MM, U.K.

Oriel Optics Corporation,
1 Market Street,
Stanford,
Connecticut 06902, U.S.A.

United Technical Supplies Ltd.,
29 Tottenham Court Road,
London W1, U.K.

Ealing Scientific Ltd.,
23 Leman Street,
London E1, U.K.

A. Jaegers (Optics),
691 Merrick Road,
Lynbrook,
New York 11562, U.S.A.

The Precision Tool and Instrument Co. Ltd.,
353 Bensham Lane,
Thornton Heath,
Surrey CR4 7ER, U.K.

12 *Lenses made to order*

Scientific Optics Ltd.,
Pondswood Industrial Estate,
St. Leonards-on-Sea,
Sussex, U.K.

12a *Meshes*
Ernest Fullam Inc.

Locker Wire Weavers Ltd.,
Woolwich Road,
London SE7 7RS
England

13 *Molybdenum trioxide crystals*

BDH Chemicals Ltd.

Hopkins and Williams

Koch-Light Laboratories Ltd.

Standard chemical suppliers

14 *Optical diffractometers for electron microscopy*

Polaron Instruments Ltd.

Rank Precision Industries
P.O. Box 36,
Guthlaxton Street,
Leicester LE2 0SP
England

Talbot diffractometer

Talbot Research
P.O. Box 41350,
Craighall,
Johannesburg, South Africa

or

AEI Scientific Apparatus Ltd.,
P.O. Box No. 1,
Harlow,
Essex, U.K.

14a *Pantograph controlled drill and punch*

Rank Precision Industries,
200 Harehills Lane,
Leeds 8, England

15 *Pinhole apertures*

Sir Howard Grubb Parsons and Co. Ltd.

16 *Polaroid films and camera backs etc.*

Polaroid Ltd.,
Welham Green,
Hatfield,
Hertfordshire, U.K.

17 *Polystyrene latex spheres*

25 ml lots in eight sizes
(0.088 μm–1.17 μm diameter)

Dow Chemical Company,
Midland,
Michigan, U.S.A.

Three convenient diameters in 8 ml lots

Ernest Fullam Ltd.

0.109, 0.312 and 0.500 μm diameter in 5 ml lots

Polaron Equipment Ltd.

18 *Thallous chloride*

BDH Chemicals Ltd.

Koch-light Laboratories Ltd.

Standard chemical suppliers

Thallous chloride is a TOXIC material

19 *Tungsten wire*

Ernest Fullam Ltd.

Johnson, Matthey and Co. Ltd.

Ladd Research Industries Ltd.

Lamp Metals Ltd.,
East Lane,
Wembley,
Middlesex, U.K.

The Tungsten Manufacturing Co. Ltd.,
Fishergate Works,
Portslade,
Brighton,
Sussex, U.K.

Polaron Equipment Ltd.

Subject index